The WRIGHT STUFF

The Century of Effort Behind Your Ticket to Space

Derek Webber

With a Foreword by
Dr Buzz Aldrin

The Wright Stuff
ISBN 9781-926592-17-6 - ISSN 1496-6921
©2010 Apogee Books/Derek Webber

Published by Apogee Books an imprint of Collector's Guide Publishing Inc., Box 62034, Burlington, Ontario, Canada, L7R 4K2
http://www.apogeespacebooks.com
Cover: Robert Godwin/Printed and bound in Canada

CONTENTS

For Joyce

FOREWORD

As I write this, it has been over 40 years since I walked on the Moon, 50 years since the start of the space era, and just over a century since the Wright Brothers' first flight. The achievements in the first century since we left the surface of the Earth have been extraordinary. We have good reasons for wanting to push new frontiers in space. We know that in the extremely long term the human race on planet Earth will need a backup plan involving space colonization (if only because of the natural development of the lifecycle of our sun). The tough part is addressing these very long-term exploration needs within the budgets of any given nation in a near-term perspective. We have made a good beginning – of all the life forms that have inhabited the planet since the dawn of time, humans have developed the means of getting off it! It was expensive and risky, and it took a great deal of creativity across the board, but we did it. So what do we do next? NASA is going to focus on developing key cutting-edge technologies to take us further, faster, and meanwhile will use commercial suppliers to deliver cargoes and travelers to the ISS. This will make it possible for NASA and other space agencies to send humans to Mars and other exciting destinations within the next two decades, and help us continue to use space exploration to drive prosperity and innovation right here on Earth.

I am excited to think that the development of commercial capabilities to send humans into low earth orbit will likely result in so many more earthlings being able to experience the transformative power of spaceflight. I can personally attest to the fact that the experience results in a different perspective on life on Earth, and on our future as a species. Space tourism is an important next step, and the way to make space traveling less expensive and less risky, and that is why I have been a strong supporter of the idea from its beginning. As this book is being published, only about 500 people have, like me, had the wonderful experience of seeing the Earth from space. That number will be exceeded in the first year of the space tourism operators (like Virgin Galactic) as they begin a new phase of spaceflight in the second decade of the 21st century. More flights on a daily basis will lead to a more airline-like operation of spacecraft.

This book is about the promise of space tourism and how we got to this point. It is written by someone who has spent over a decade working on the business cases and the regulations to make it viable. Derek helped me to present the case for space tourism when I was a Commissioner on the President's Commission on the Future of the US Aerospace Industry. He illuminates an intriguing tale of the historical threads and people that have come together to make space tourism possible.

Buzz Aldrin
Los Angeles, California
March 2010

INTRODUCTION

"We need affordable space travel to inspire our youth, to let them know that they can experience their dreams, can set significant goals and be in a position to lead all of us to future progress in exploration, discovery and fun."

Burt Rutan, SpaceShipOne designer, 2004

So, you are about to become a space tourist. That's great. Now you want to know something about how this new industry of space tourism came about. This book tells the story of what it took to make space tourism an overnight success. It involves a mix of types of people from a range of nationalities over a hundred years. They include scientists, engineers, aviators, politicians, artists and business executives. Several of the individuals belong in more than one category. We are going to have some fun presenting awards as we track the story. As you probably know, space tourism has already begun, at least in the form of the expensive orbital experience, using Russian vehicles. Sub-orbital space tourism flights like yours are just about to begin, will be significantly less expensive, and will operate at least initially from the US. There are long waiting lists, and it is expected that around 10,000 people like you each year will be opting for the opportunity once the operational flights become established. It took 100 years to get here. As you are about to become a space tourist, you stand next in line to an amazing historical tradition, which this book will reveal to you. Well done for taking that next step! In some respects, what follows is a parallel history of aviation and space travel.

This is also a book about linkage, threads and connections. When we look back at the development of this space tourism business, we discover several historical threads, each one of which is important and interesting in itself, but which became very powerful when operating as an intertwined whole, much as a rope derives strength from its constituent fibers. The artists helped transfer the vision of the scientists and engineers to the public, whose voice was heard by the politicians. Finally, the business people came on the scene, to make space travel available for all.

We shall see that certain individuals have been able to link the very beginnings of flight with the personal spaceflight industry now taking shape. Some of the linkages may surprise you. In particular, we find that there has been an interconnectedness between the fields of aviation and rocketry, from their respective beginnings. Space tourism represents the ultimate coming together of two distinct fields: aviation and rocketry.

The Wright Brothers, without formal training, were very thorough scientists as well as being engineers. Charles Lindbergh, it emerges, was a link between those earliest times of flight and space tourism today. Wernher von Braun was an engineer and a marketing man, and members of his team have helped until today with the development of personal space flight operations. According to Tom Wolfe's book "The Right Stuff", the test pilots who had been pushing back the boundaries of high speed flight in the high atmosphere at Edwards Air Force Base (the X-men), seemed to be sidelined by the arrival of the Project Mercury astronauts. Yet the real story is not so simple. And Burt Rutan, the designer of SpaceShipOne, the first civil space vehicle, learned his flight test engineering skills at Edwards. Again, we see the connections through time that have led to today's space tourism operations.

Aviation was of interest to the military and politicians from its onset. And subsequently, the space program has been a matter of interest to all the American presidents since Truman, and they have had to deal with the political ramifications of the change in focus of the program over the last half-century, reflecting the changing attitudes of the American public. As you might well imagine, politicians have taken advantage at times of triumph, and often retreated when things were more difficult. The administration of one president did not get much credit for doing so, but it put in place the enabling legislation and regulation for an emerging space tourism industry. This was a politically risky thing to do. As you fully realize, and despite the care of the operators, there exists the real possibility of accidents to rich, and potentially influential, space tourists in this program. America, at its core, is about accepting risk in order to progress. And space tourism is certainly a worthy exemplar of that notion. This outcome (of creating a civilian space program) was not by any means inevitable, and only happened because of the committed efforts of a number of individuals. And artists have also had an important role from the very beginning in bringing the experience and possibilities of air and space travel to a wide public.

So, what was, and is, the "Wright Stuff", this ingredient or mix that enables everyman to progress to a future in space? It is a combination of things. First there must be vision, then a certain thoughtful acceptance of risk in order to achieve great things in the air and space. There must be a scientific and engineering appreciation of the laws of nature, and how to use them to advantage. Ultimately the marketing and financial know-how is needed to be able to give the public what they want, while making a buck. Oh, and maybe we should add into the mix the final essential ingredient: enthusiasm. Mix together in a supportive political environment, and you get space tourism! Our motto is: "Look before you leap, but then you must leap!" It's no good if you forget the "leap" part. That does not get us anywhere. We shall make "Wright Stuff" Awards to deserving recipients, who have indeed leapt, as we progress through the book.

The fascinating thing about this tale is that it all took place since the advent of photography, and so we can see the players over time. It also took place almost within living memory of the oldest of us today. The author himself was around for over half of the events described, either simply as an interested citizen and aviation enthusiast, or later as a professional involved in the commercial aerospace business and the attempts to create a new industry of space tourism. This personal experience and enthusiasm of the author, partly as chronicler and partly as a minor participant, has to

some extent guided the selection of the material included. Hopefully, some of that enthusiasm will be transmitted to the reader. These have been extraordinary times. We can see through the photographs something of how the key players interacted. In some respects what emerges seems like a kind of relay race, where each generation pushes back on the boundaries, going faster and higher, before handing over to the next generation. However, that is a view colored by hindsight, which implies a kind of inevitability about it all. This book aims to show that progress depended not so much on subsequent generations, but on visionary *individuals* in each generation. These folk were the real torchbearers, who carried the flame between the generations, reaching both backwards and forwards through time and often across categories. It is only in retrospect that we can see how important were their contributions and risk taking. We learn that risk takes many forms. There is of course physical risk to life and limb, but also political, business and career risks have been necessary to the development of spaceflight for all. Some individuals demonstrated a willingness to take on more than one of them at a time.

Enjoy looking at these photographs, and following the threads of history that have led us to this point, the point of the emergence of a space tourism industry, when you can have your very own spaceflight experience. We shall follow the threads in turn of the aviators, the rocket men, the X-plane pilots, presidents, artists and space tourism pioneers who made this possible. Some special individuals belong in more than one of the categories and we shall therefore see their names and achievements pop up in multiple chapters of the book. We may find ourselves talking about rocketry, for example, in the middle of a chapter on aviation. That's just the way it has been in the real world, so just be prepared to roll with these diversions, and appreciate the intertwining of the story as it unfolds. More information about the key individuals whose connected story is told in these pages is provided in the side bar insets, and in the books of the Bibliography.

Space tourism, it turns out, is not a sideline or aberration in the mainline of space developments, but is the latest and enabling stage in a historical chain of developments that can be traced back to the Wright Brothers, a century ago.

YOUR TICKET TO SPACE

"Poyekhali!" ("Let's go!")
Yuri Gagarin, Baikonur, 12thApril, 1961

You are about to take a ride into space, and you want to know more about the historical chain of events, and the extraordinary key people whose efforts, and in many cases personal courage, have made this possible. Table 2-1 provides a summary of all commercial space traveler flights to date. They have all been orbital flights, and have all been by means of a Soyuz rocket launch from the former Soviet spaceport of Baikonur, which is now in Kazakhstan. Orbital tourism flights circle the Earth once every 90 minutes, and the traveler is weightless for maybe up to two weeks during which the astronaut sees 16 sunrises and sunsets each day. Orbital tourists generally rendezvous with somewhere more spacious than their spacecraft, whether it is a space station, or eventually a space hotel. The small total number of commercial manned orbital flights to date is because of the relatively high cost of orbital launches, together with limited availability of tourist seats in the Russian Soyuz spacecraft. Soyuz carries only thee people, and usually at least two of them are professional government cosmonauts or astronauts, so it has been a rare and special opportunity when, not generally more than once a year, a place could be made available for space tourists. Only nine private civilian space travelers have been able to experience space in the half century since the start of the space program (with one of them flying twice). In that time approximately 500 government astronauts have flown. All of this will, as you know, change very rapidly as the sub-orbital space tourism business begins to operate, at one hundredth of the cost of the orbital trips. But these nine citizens have blazed the way for the rest of us. We shall honor them individually later in the book.

TABLE 2-1 COMMERCIAL SPACE TRAVELER FLIGHTS AND SPACECRAFT

LAUNCH DATE	TRAVELER	NATIONALITY	VEH/SPACECRAFT DESTINATION	SELF-FUNDED?
2nd Dec 1990	Toyohiro Akiyama	Japan	Soyuz/Mir	No
18th May 1991	Helen Sharman	UK	Soyuz/Mir	No
28th April 2001	Dennis Tito	US	Soyuz/ISS	Yes
25th April 2002	Mark Shuttleworth	South Africa	Soyuz/ISS	Yes
1st Oct 2005	Greg Olsen	US	Soyuz/ISS	Yes
18th Sept 2006	Anousheh Ansari	US/Iran	Soyuz/ISS	Yes
7th April 2007	Charles Simonyi (A Flight)	US/Hungary	Soyuz/ISS	Yes
12th Oct 2008	Richard Garriott	US/UK	Soyuz/ISS	Yes
26th March 2009	Charles Simonyi (B Flight)	US/Hungary	Soyuz/ISS	Yes
30th Sept 2009	Guy Laliberté	Canada	Soyuz/ISS	Yes

You are probably going on a sub-orbital trip from one of the new spaceports that have been established to provide the space tourism experience. Spaceports like Mojave in California and Spaceport America in New Mexico, or maybe Kiruna in Sweden. Sub-orbital trips go straight up to the 62 mile (100km) notional boundary of space, the passengers see the Earth's curvature, the thin veil of the atmosphere which ensures our survival, they experience maybe 5 minutes of weightlessness, then return straight back down to the spaceport without going into orbit.

Chapter 8 will tell you about the people who have worked since about 1990 to bring the new industry of space tourism into existence.

Chapters 3 thru 7 go back even farther, right to the beginnings of aviation and rocketry in 1903, to point out the important historical connections that built the foundations on which the space tourism industry could be established. Your flight would not have been possible without the creativity, efforts and risk-taking of these individuals. Fig 2-1 captures the dawn of the space tourism era. The author took the photo just as the sun was emerging above the distant desert horizon at Mojave Spaceport, California, in 2004. Tens of thousands of the public had managed to find their way to this isolated spot to watch history being made, as SpaceShipOne, the first civilian space ship, roared off into space as an entrant in the Ansari X-Prize, about which you will learn later.

Fig 2-1 X-Prize Dawn, 29th September 2004, Mojave, California.

And what are the views you can expect to see as a space tourist? The orbital tourist can see just about the entire surface of the Earth through successive orbits, and the sub-orbital tourist can see over 700 miles in all directions from above the spaceport of launch. Fig 2-2 provides a fine example showing the Great Lakes from an orbital mission.

Fig 2-2 The space tourist's view from orbit. The entire Great Lakes seen from a Space Shuttle.

(Credit: NASA)

And by the way, don't be concerned about what people call you. Some folks use the term "space tourist". Others use various combinations of the following: space flight participant, public space traveler, private space adventurer, private space explorer, personal spaceflight participant, private astronaut, civilian astronaut, personal space traveler. The author will generally stick with "space tourist" for the purposes of this book. Whatever name you give yourself, you are about to do something that is pretty extraordinary. So, have a great ride. Meanwhile, we are going to find out how this opportunity came about.

We begin by looking at the developments in aviation that were necessary. In order for space tourism to eventually be possible, we have needed to advance in a series of steps from the first flight, to long distance flight, to passenger airlines, to jet flight, and to supersonic passenger flight. Chapter 3 takes us on that journey. The first aircraft passengers on scheduled flights were generally wealthy, and somewhat adventurous, individuals. But, as aviation developed the combined reliability improvements and passenger economics have resulted in today's situation where just about anyone who wants to fly, can fly. This is a pattern that will surely be followed by the space tourism business. If your friends cannot afford a ticket now, then they will be able to, eventually

How can they know the joy to be alive
Who have not flown?
To loop and spin and roll and climb and dive,
The very sky one's own,
The urge of power while the engines race,
The sting of speed,
The rude wind's buffet on one's face,
To live indeed

B.P.Young
(from "Flight", quoted in "Icarus: An Anthology of the Poetry of Flight", 1938)

Pilots or flyers - that's what they are called today. But in the early years, they were called aviators. It somehow conveyed much more than the word pilot ever could. Back then, they often designed and built their aircraft too, so the original word somehow conveys that fact, as well as the associated daring of venturing forth into the relatively unknown and dangerous realm of the air. You could ask ordinary people, whom you met in the street, and they all knew the names of the aviators. Both the aviators, and the names and the design of their aircraft, were famous, and were the subject of countless news stories. The most famous aircraft from the early years of aviation were probably the Wright Flyer of 1903 (Fig 3-1), The Bleriot XI Monoplane of 1909 (Fig 3-2), and the "Spirit of St Louis" of 1927 (Fig 3-3). Back then, they were called "flying machines" and later "aeroplanes". These three craft, all made of a simple framework covered with canvas, illustrate the dramatic steps that were taken by the early pioneers in a relatively short timeframe. They all had courageous aviators who risked their lives to nurse their craft into the air. We all benefited.

Fig 3-1 The Wright Flyer of 1903 – now on display at the Smithsonian Institution's National Air and Space Museum in Washington, DC. It flew four times, on December 17th, and never flew again.

(Credit: Author)

Fig 3-2 Bleriot XI Monoplane of 1909. This example is at the Owlshead Museum of Transportation in Maine.

(Credit: Author)

Fig 3-3 The "Spirit of St Louis" of 1927, at the National Air and Space Museum, Washington, DC.

(Credit: Author)

The Wright Brothers, Orville and Wilbur (one should say the *younger* Wright Brothers, because there were also two older brothers as well as a sister) (Fig 3-4), had been carrying out a series of scientific and engineering experiments and test flights starting around 1900 at Kitty Hawk in North Carolina. Even before they began their practical experiments, they studied the theoretical work of predecessors such as Otto Lilienthal, S.P. Langley, Sir George Cayley, Octave Chanute, and even Leonardo da Vinci (see Ref 1, Ref 6). In England, Sir George Cayley, in 1799, had drawn a design for a practical aircraft. During the last decade of the 19th Century, Otto Lilienthal in Germany had conducted about 1,000 manned glider flights, before his fatal accident. And Percy Pilcher in Britain was also active until he too lost his life in a crash. Chanute had been a fine chronicler and correspondent of all the experimenters around the world and he wrote a definitive consolidation book in 1894 called "Progress in Flying Machines". And there had of course been successes with lighter than air flight a century earlier – the Montgolfier brothers flew one of their balloons with two passengers in 1783 for 25 minutes covering 5 miles. All of this was important background for the Wright Brothers as they conducted their experiments.

The original aim of the Wright Brothers was scientific, and not financial, although they later became tied up in patent battles and were able to make money from their invention. They endured a great many tribulations, both due to nature and due to their fellow men, but always with good grace, charm and modesty. They were only in their early thirties when they achieved their breakthrough, and we will notice as we follow this history that often the breakthroughs happen at the hands of men and women who are in their twenties or thirties. The Wrights were careful researchers, and persistent in adversity. And they were very well aware of the risks they were taking, having studied the fatal flights of Lilienthal and others for clues. They wrote back home to their father: "If you are looking for perfect safety, you will do well to sit on a fence and watch the birds; but if you really wish to learn, you must mount a machine and become acquainted with its tricks by actual trial." They not only designed and built their aircraft, but they designed and built their engine and propellers too. They practiced first with gliders until they understood the control processes, then introduced an engine.

Name Wilbur and Orville Wright

(Credit: Science Museum – Science and Society Picture Library)

Summary Description
Pioneers of Flight (Dec 17th, 1903)

Date of Birth
Wilbur April 16 1867, Millville, Indiana, USA
Orville August 19 1871, Dayton, Ohio, USA

Date of Death
Wilbur May 30 1912, Dayton, Ohio, USA
Orville January 30 1948, Dayton, Ohio, USA

Nationality US

Achievements
Early aerofoil designs (used wind tunnel for testing). First Flight of controlled powered heavier than air machine

Specific help for Your Ticket to Space
Until the Wright Brothers were able to move from theory into practice in 1903, no-one knew for sure that man could make a controlled flight in a heavier-than-air machine. And of course, once they knew it could be done, others wanted the experience too. Everyone agreed that the experience of seeing the ground from aloft for the first time was a transforming event. Unfortunately it was a rather dangerous endeavor in the early years, but nevertheless this was something worth the risk. Back then, there were no government regulators to prevent the early test flights, and progress was quick, through trial and error, and through competition. Nowadays the government regulators of the new space tourism industry (the US FAA-AST) have tried to keep largely "hands-off", while the industry becomes established, emulating the conditions that led to success for the Wright Brothers. So, we owe a great deal to those who started it all. And of course, without the Wright Brothers, we could not have the "Wright Stuff" Awards.

Theirs was the first practical powered flight in a controllable heavier than air machine, achieved on December 17th 1903. So, this is the starting point we have chosen for the story of your ticket to space. The Wrights chose a bleak stretch of coast in North Carolina called Kitty Hawk to carry out their trials, because of the steady winds and soft sandy terrain in the event of accidents. They persevered through three years often in atrocious weather, living in tented accommodation (later converted to wooden sheds), until they achieved their success. The author visited the site almost a century later. It was a bleak day in 2001, and it was very easy to imagine the hardships the Wrights had suffered. Today there is a memorial marker indicating the beginning and end of each of their flights that day. And four flights were flown, each one going farther. The first flight (Orville was the pilot) went for 12 seconds and covered 120 feet. The fourth flight (with Wilbur at the controls) lasted 59 seconds and covered 852 feet. The aircraft never flew again. It was blown over in a gust of wind, and the damaged parts crated up and returned to their hometown of Dayton, Ohio. The Wrights were not good at P.R., and the news of their flight did not immediately make the headlines. Many newspaper publishers were unable to understand the distinctive difference of their achievement compared with balloon flights, which were by then quite common. The cover of this book shows the front page from one of the few newspapers that recorded the event – two days later - it was the *Cincinnati Enquirer*. The Wright Flyer has subsequently been repaired to its original state, and now is in the Smithsonian, but the Wrights built improved versions through to 1909 while attempting to win a US government contract. So much of future progress depended on the careful and courageous work of the two men who designed and flew their frail craft into the high winds at this bleak Kitty Hawk flying field. If you have never been there, the journey to Kitty Hawk is recommended as a modern pilgrimage to all future space tourists, such as yourself.

No one would argue that The Wright Flyer looks anything like any of today's aircraft, but it was nevertheless the result of a rigorous process of development and testing. The Wright Bros used a homemade wind tunnel in order to validate their assumptions. And of course the machine worked.

The human pilot was able to control it. "The problem of flight" as it used to be called, had been solved.

Fig 3-4 The Wright Brothers, Orville and Wilbur.
(Credit: Community College of Denver)

As they had written to their father, what the Wright Brothers were doing was not without risk. They had accidents and crashes, including the death of a passenger, while trying to interest the US government in buying their machine. They had greater success, however, in Europe, where various princes, industrialists and newspaper moguls came to watch Wilbur perform. Among those watching Wilbur Wright demonstrating controlled flight, flying a circle at le Mans in France in the summer of 1908, was Louis Bleriot. He declared: "For us in France and everywhere, a new era in mechanical flight has commenced….It is marvelous." (Ref 6). The international press at last recognized the Wrights' achievement from 5 years earlier. They would not however stay in the lead for much longer.

In France, a number of the early aviators in designing their own machines were using the data about control techniques published by Chanute about the Wrights. Louis Bleriot was a quick learner. Look at the Bleriot machine in Fig 3-2. It is totally recognizable as a modern aeroplane.

First of all it has its elevator control surfaces (for making the craft go up or down) at the rear of the craft, not in front as was the case with the Wright Flyer. And of course it was a monoplane. In this craft, Louis Bleriot took the risky and courageous challenge of flying from France to England across the 22 miles of the English Channel. This was only 6 years after the start of aviation with the Wright Brothers at Kitty Hawk (see Fig 3-5 and Fig 3-6), and only a year after watching Wilbur Wright demonstrate the control technology. In that short time interval, the level of confidence in the engine and technology had reached the point where such a venture was possible. Bleriot won a 1,000 pounds Daily Mail prize for his feat.

Fig 3-5 Bleriot leaves Calais heading for Dover, Aug 3, 1909.

(Credit: Aviationhistory.info)

Fig 3-6 Bleriot arrives at Dover (he is facing the camera, in flying gear).

(Credit: Aviationhistory.info)

Imagine the impact this had on the island nation of Great Britain with its proud naval tradition. The newspapers went crazy. The fascination with aviation was also a global phenomenon. The US, Germany, France, and Britain were all involved from the beginning; and the Brazilian Alberto Santos-Dumont also had a successful monoplane at that time.

Aviation was transforming the international transportation, political, and military scene. During the First World War (1914-1918), aviation first entered the military arena. This hastened developments, so that by the end of the war aircraft had become faster and more reliable. There was a surplus of machines after the war effort, and aircraft were made available in large numbers for private purchase, for training purposes and for airmail. The US Airmail started in 1918 flown by government pilots in Curtis Jennies. Then the 1925 Kelly Airmail Act ensured that private contractors could thereafter carry airmail. It was a dangerous business. Of more than 200 pilots hired by the Post Office from 1918 to 1926, 35 died flying the mail (Ref 5). Charles Lindbergh was a product of this era and became an airmail pilot. He flew the first air mail flight from Chicago to St Louis in April 1926. It was to the businessmen of St Louis that Lindbergh turned for funding for his next and grand venture.

There were many prizes for feats of aviation in those early years, often funded by newspapers whose aim was to increase circulation, because the general public was captivated by aviation from its onset. We already mentioned Bleriot's prize. But the person who earned the most attention from the public was unquestionably the "Lone Eagle", Charles Lindbergh, who flew his "Spirit of St Louis" from New York to Paris, on 20-21 May 1927, thereby becoming the first person to fly solo non-stop across the Atlantic (Ref 2, Ref 7). Fig 3-7 introduces Lindbergh in his cockpit. Although only 25, Lindbergh was a

very experienced flyer and excellent navigator. In an era before navigational aids, aviation was very much a matter of awareness of weather and geography. Lindbergh had even had to bail out on three occasions due to engine and fuel problems with his earlier aircraft, when he was a National Guardsman and a mail pilot, and when there was simply no alternative in order to return to earth safely. He was taciturn in temperament, and a very careful planner. He named the plane for the city of his financial backers, even though it had been built in San Diego, and was to fly to Paris from New York. He loved flying. In his book (Ref 2) he says: "Science, freedom, beauty, adventure: what more could you ask of life? Aviation combined all the elements I loved."

Fig 3-7 Charles Lindbergh in the cockpit of the "Spirit of St Louis". He was only 25 when he undertook his historic flight.

(Credit: Lindbergh Picture Collection, Manuscripts and Archives, Yale University Library)

This New York to Paris flight, which took 33 ½ hours, happened in 1927, only 18 years after Bleriot's feat. Lindbergh was in a competition to try to win the $25,000 Orteig prize for being the first to fly non-stop from New York to Paris. Many more famous aviators tried to win the prize and failed. However, it is generally recognized that the success of Lindbergh in flying the Atlantic was the trigger that started the whole modern airline business, the next essential step in eventually getting us all into space.

Lindbergh became famous overnight, it was estimated that 4 million people came out to greet him in New York (Fig 3-8), and his achievement was celebrated all around the world. Fig 3-9 is a later commemorative postage stamp. We shall see that this fame became a very important factor in the subsequent development of what became the space tourism business. But that was much later.

Fig 3-8 Reception in New York City for Charles Lindbergh, June 13th 1927.

(Credit: Lindbergh Picture Collection, Manuscripts and Archives, Yale University Library)

Fig 3-9 Lindbergh commemorative postage stamp.

Name Charles Augustus Lindbergh

(Credit: Lindbergh Picture Collection, Manuscripts and Archives, Yale University Library)

Summary Description   Aviator.

Date of Birth
4 February 1902, Detroit, Michigan, USA

Date of Death
26 August 1974, Maui, Hawaii, USA

Nationality US

Achievements
Designed and flew "Spirit of St Louis", the first solo non-stop transatlantic flight, 1927.
Business associate of Juan Trippe, helping establish the PanAm international airline route network.

Specific help for Your Ticket to Space
In many ways a modern space tourism vehicle represents the coming together of the two distinct fields of aviation and rocketry. Lindbergh epitomizes this coming together, because of his pioneering aviation achievements and his support of Robert Goddard's early rocketry experiments. He was also an early participant in the Mayo Clinic's research into the physiological limits of high speed and high altitude flight. Charles Lindbergh was only 25 when he flew the Atlantic solo to win his accolades. Next time you are half way across the Atlantic at 30,000 feet, reclining in your seat and watching a boring movie, just imagine the Lone Eagle in his cold and noisy cockpit, with no direct forward vision, traveling the route at 100 mph and a few hundred feet altitude for 33 hours while remaining awake. That's the kind of spirit that was needed to get this business started. Are you 25 yet?

Figures 3-10 and 3-11 let us share some early recognition for Lindbergh, following his Trans-Atlantic flight. Lindbergh met both Orville Wright and Louis Bleriot, neatly linking together for us those first 24 years of aviation (Wilbur had died 15 years earlier of typhoid fever contracted on one of his international trips to defend his design patents). Lindbergh, we should realize, had been only one year old when the Wright Brothers opened up the world of controlled, powered, heavier-than-air flight.

Fig 3-10 Orville Wright meets Charles Lindbergh after the Paris flight, on June 22nd 1927, in Dayton, Ohio.

(Credit: Corbis)

However, Lindbergh was not only a link to the past. He was a very important link to the future and indeed right up to our day. Figure 3-12 introduces the next key figure in the developing story. Lindbergh, whose father was a congressman, was introduced to the wealthy Harry Guggenheim, who had been captivated by the aviator's exploits. Guggenheim represented the Daniel and Florence Guggenheim Foundation, and was an early aviation supporter. He had visited Lindbergh before he left for Paris, saying, "Look me up on your return", not being at all certain that Lindbergh would survive the venture. On his return, the two men resumed contact, and Guggenheim sponsored a 3-month nation-wide tour for Lindbergh and the "Spirit of St Louis"- 48 states, 92 cities and 147 speeches. The tour map is provided in Fig 3-13. It is estimated that *a quarter of all Americans* personally saw Lindbergh on this tour. The public thereafter, as a consequence, began to see airplanes as a viable and reliable means of travel and for delivering long distance mail. The "Spirit of St Louis" went on to fly a Goodwill Tour of the countries of Central America and the Caribbean before ending its days at the Smithsonian's National Air and Space Museum in May 1928.

Fig 3-11 Lindbergh is greeted by Louis Bleriot, June 3rd, 1927, Paris.

(Credit: Corbis)

Fig 3-12 Lindbergh and Harry Guggenheim in 1927.

(Credit: Lindbergh Picture Collection, Manuscripts and Archives, Yale University Library).

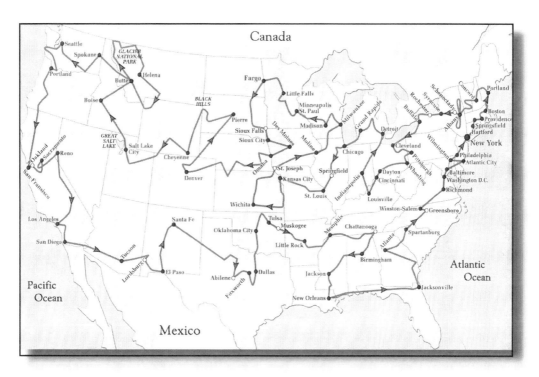

Fig 3-13 Lindbergh's Guggenheim tour map.

Fig 3-14 Saint Exupery, French airmail pilot and author.

Fig 3-15 Airmail celebratory postage stamp.

The airmail pioneers, such as Lindbergh himself, came first, locating emergency landing fields and fuel depots, and mountain passes by which the early aircraft could steer through mountain ranges; passenger air travel followed after the routes were established.

Many of those who opened up the early air routes would die in the attempt. The famous French aviator and author Antoine de St Exupery (Fig 3-14) (he wrote *"Courrier Sud, "Vol de Nuit", Le Petit Prince"*) was an example of an airmail pilot who pioneered air routes in South America (see Ref 3). He, like the Pulitzer Prize-winning Lindbergh, spread the notion of "the romance of the air" through his almost poetic writing to an eager public.

Applications for pilot's licenses in the US in 1927 increased by 300%, and US airline passengers increased from around 6,000 in the year before Lindbergh's pioneering flight to over 170,000 in the year after his flight. Fig 3-15 notes a later celebration of the airmail service on a postage stamp. Some have described the '20s and '30s as a Golden Age of aviation, and there was certainly a constant supply of record attempts and air races to maintain the public's interest during that era (see Ref 5).

The first air passenger was carried by one of the Wright brothers, sitting on the wing of their Flyer, but the first true airline companies began flying a few passengers for short distances in around 1914. The Dutch airline KLM was founded in 1919 and is the oldest established airline that is still operating. Flying was not comfortable and there was a great deal of vibration, turbulence, noise and the all-pervading smell of gasoline and engine oil. It was very cold particularly in the winter months, and there was no air conditioning for the summer. We need to acknowledge and thank these wealthy early air travelers, who had to endure such trials and tribulations, because they made it possible for the rest of us to fly later. Passenger flying started out as a premium business, and only later was it possible to bring the price down so that everyone could fly. The early passengers provided the revenue-based funding which enabled the first airlines to begin to operate, and to encourage new aircraft designs to be developed. This is likely to be the case with space tourism, too. Fig 3-16 lets us experience such an early flight, although the noise and vibration are not apparent; the photograph having probably been taken on the ground. We know that the early air hostesses (as flight attendants used to be called back then) had to use megaphones in order to be heard by the passengers. Fig 3-17 is a reminder that weight was at a premium, and the first passengers had to be weighed before being allowed to board their aircraft (in this case it was at London's Croydon aerodrome, before a November 1934 flight to Scotland).

Fig 3-16 An early passenger flight – note the wicker seats.

Fig 3-17 Early passengers were weighed before flight (1934).

(Credit: Science Museum – Science and Society Picture Library)

Other factors, than the simple need to get from A to B, also motivated the early airline business. The Aeromarine airline, for instance, operated modified Curtis Flying Boats between Miami and the Bahamas and Cuba through the years 1920-1924, effectively skirting the US prohibition era restrictions about consumption of alcohol!

However, the large growth in passenger air travel, which followed Lindbergh's 1927 flight, had generally a rather more noble purpose, and was to a considerable extent a consequence of Lindbergh's relationship with business entrepreneur Juan Trippe (see Fig 3-18 and Fig 3-19). Trippe established Pan American, flew its first flight on October 19, 1927, from Key West, Florida to Havana, Cuba, and thereafter introduced truly global passenger air travel, and conceived of the Tourist Class. Lindbergh surveyed the routes for Trippe's operations. He flew a major survey trip to the Orient in 1931, and around the North Atlantic in 1933, with his wife Anne Morrow Lindbergh as co-pilot, navigator and radio operator. We can recall the social situation of women in that era by recounting the tale of a journalist who interviewed Lindbergh before taking off with his wife in their small plane

on one of these expeditions. "Don't you think it is dangerous taking a lady on a flight like this?" said the journalist "I'm sure I would not dream of taking my wife". Lindbergh's reply was brief, pointed, and showed his appreciation for women: "You do realize", he said "that she's crew?"

In 1933, Lindbergh wrote far-sightedly to Trippe: "I believe that in the future aircraft will detour bad weather areas by flying *above* them rather than around them". He was certainly looking very far ahead. During the Second World War, Lindbergh performed war service flying sorties in Corsairs and Lightnings in the Western Pacific during 1944.

Fig 3-18 Juan Trippe and Charles Lindbergh in Panama in 1929 during a Fokker F10A survey flight.

(Credit: CharlesLindbergh.com)

Fig 3-19 Juan Trippe and Charles Lindbergh and an early Pan American Sikorsky S-38 flying boat, September 1929, British Guiana.

(Credit: Lindbergh Picture Collection, Manuscripts and Archives, Yale University Library)

Name Juan Terry Trippe

(Credit: University of Miami Libraries)

Summary Description
Aviation entrepreneur, Founder of Pan American Airways

Date of Birth
June 27 1899, Sea Bright, New Jersey, USA

Date of Death
April 3 1981, Los Angeles, California, USA

Nationality US (he was named after his Cuban grandfather)

Achievements
Created Pan American Airways.
Major force behind the Boeing 707 and 747.

Specific help for Your Ticket to Space
Juan Trippe created the international air travel business. Before he came along, flying was largely seen as either: a military thing, a postal delivery service, or the domain of rich and/or crazy adventurers. Only some short distance air routes had started, taking a few of the elite. From the very beginning, he wanted to bring aviation to the general public, and he was the first to create the tourist class with that aim in mind. In hindsight, perhaps some of us who travel in today's crowded tourist class (sometimes referred to as cattle class) might have cause to question whether the outcome is what Trippe intended. However, we must acknowledge our debt to his vision and his practical zeal in bringing about that vision. We have now all become air tourists, in one way or another, and so this makes the next step of becoming space tourists that much closer.

After the Second World War (1939-1945), the United Nations was formed to provide a forum to make it possible for international dialog to take the place of war. International travel was also an important contribution to the development of global understanding and global trade. In June 1947, Juan Trippe and his wife, and a cabin full of VIPs, made an inaugural commercial round-the-world flight of PanAmerican Airways in a Lockheed Constellation. We shall come back to Juan Trippe and his ongoing contributions later. Lindbergh also worked as a consultant for TWA, which for a while was known as "The Lindbergh Line".

But now we need to take a slight diversion for something completely different - which simply will not wait until the next Chapter. Recall that I did warn you at the outset that some of the players in this story refuse to be placed in a single category. Around the same time that Lindbergh was planning his transatlantic flight in 1926, in Massachusetts, Robert Hutchings Goddard (Fig 3-20), a physics professor at Clark University in Worcester, was beginning to design and launch small liquid propelled rockets. His first successful launch of a liquid-fueled rocket took place on March 16th, 1926, in Auburn, Massachusetts, and the vehicle reached perhaps discouragingly only 40 feet in altitude. Of course, the Wrights only flew 120 feet on their first flight. So the trick is not to be easily discouraged. We shall learn more of Goddard in Chapter 4, but we should notice for now how the timescales overlap.

Goddard had published his first theoretical paper "A Method of Reaching Extreme Altitudes" as a Smithsonian pamphlet in 1920, and was just beginning to experiment when Lindbergh flew the Atlantic. Goddard wrote his second paper "Liquid-Propellant Rocket Development" in 1936 (see Ref 4). And the work to make this possible would not have happened without the intervention of Charles Lindbergh, which is why he appears as a diversion in this "Aviators" Chapter. Lindbergh had learned about the work that Goddard was doing on a shoestring with a small amount of Smithsonian funding ($1000/ year), and felt that it was a potentially important technology of the future, deserving support. And so he arranged for Goddard to get a major injection of funds ($100,000) from Guggenheim to enable him to continue his research and test flights. Thereafter, Goddard no longer needed to worry about funding and Guggenheim paid all his costs each year. Fig 3-21 reveals Lindbergh, Goddard and Guggenheim at the Roswell, New Mexico launch site, which was supported through the Guggenheim funding.

Fig 3-20 Robert Goddard with early liquid pro-pellant rocket.

(Credit: NASA)

Name Dr Robert Hutchings Goddard

(Credit: NASA)

Summary Description
Physicist and Pioneer rocket designer

Date of Birth
October 5 1882, Worcester, Massachusetts, USA

Date of Death
August 10 1945, Phoenix, Arizona, USA

Nationality US

Achievements
Designed and built liquid propelled rocket motor

Specific help for Your Ticket to Space
Goddard was a very methodical experimenter and inventor who created and test flew a range of liquid propellant rockets that attained ever higher altitudes. It is probably fair to say that he would have been astonished at the development of space tourism – because he saw his rockets as being needed primarily to conduct measurements at high altitude as part of understanding the Earth's atmosphere. His early inspiration came from novelist H.G.Wells. His dedication to his work drew the attention and support of Charles Lindbergh who had a perhaps wider vision of the potential utility of Goddard's rockets. Goddard developed the rocket equations in the US, almost simultaneously with Konstantin Tsiolkovsky in Russia, and went further than Tsiolkovsky by building and launching the resulting vehicles. All subsequent rocketry work, including the rockets used in space tourism vehicles, use these basic equations. So when folks say "…. It's like rocket science", it is to the work of Goddard that they refer.

Fig 3-21 Harry Guggenheim (2nd from left), Robert Goddard (center), Charles Lindbergh and launch crew A.W. Kisk and N.T. Ljungquist, on 23rd September, 1935 at Roswell, New Mexico.

(Credit: Lindbergh Picture Collection, Manuscripts and Archives, Yale University Library)

Goddard was able as a consequence to set up his operations in New Mexico, and from there continue the developments of his rockets right up to his death on August 10th, 1945, at the end of the Second World War. Fig 3-22 discloses a letter that Goddard wrote to Lindbergh as his sponsor in 1937, providing him with details of his latest rocket experiments.

Lindbergh's interest in rocketry did not end with Goddard. Another rocket engineer was making his presence felt around 1945, particularly if you lived in the London area. Wernher von Braun (Fig 3-23) was at that time launching V2 rockets from Peenemünde on the Baltic coast across the North Sea as part of a German blitz on London. The photo shows him after the war working on the directional fins at the tail end of a V2.

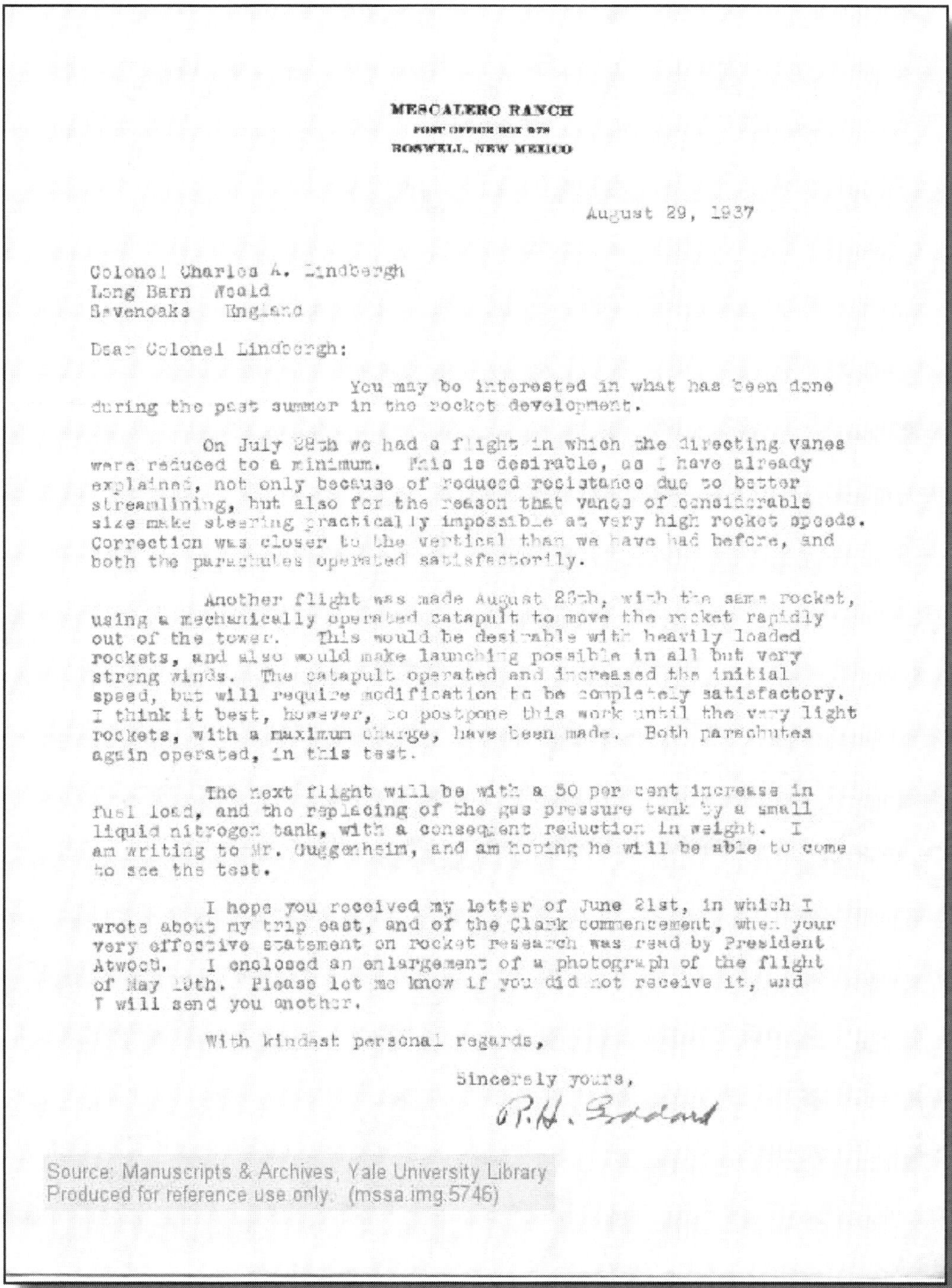

MESCALERO RANCH
POST OFFICE BOX 978
ROSWELL, NEW MEXICO

August 29, 1937

Colonel Charles A. Lindbergh
Long Barn Weald
Sevenoaks England

Dear Colonel Lindbergh:

You may be interested in what has been done during the past summer in the rocket development.

On July 28th we had a flight in which the directing vanes were reduced to a minimum. This is desirable, as I have already explained, not only because of reduced resistance due to better streamlining, but also for the reason that vanes of considerable size make steering practically impossible at very high rocket speeds. Correction was closer to the vertical than we have had before, and both the parachutes operated satisfactorily.

Another flight was made August 26th, with the same rocket, using a mechanically operated catapult to move the rocket rapidly out of the tower. This would be desirable with heavily loaded rockets, and also would make launching possible in all but very strong winds. The catapult operated and increased the initial speed, but will require modification to be completely satisfactory. I think it best, however, to postpone this work until the very light rockets, with a maximum charge, have been made. Both parachutes again operated, in this test.

The next flight will be with a 50 per cent increase in fuel load, and the replacing of the gas pressure tank by a small liquid nitrogen tank, with a consequent reduction in weight. I am writing to Mr. Guggenheim, and am hoping he will be able to come to see this test.

I hope you received my letter of June 21st, in which I wrote about my trip east, and of the Clark commencement, when your very effective statement on rocket research was read by President Atwood. I enclosed an enlargement of a photograph of the flight of May 19th. Please let me know if you did not receive it, and I will send you another.

With kindest personal regards,

Sincerely yours,

R. H. Goddard

Fig 3-22 Goddard letter to his supporter Charles Lindbergh in August 1937.

(Credit: Lindbergh Picture Collection, Manuscripts and Archives, Yale University Library).

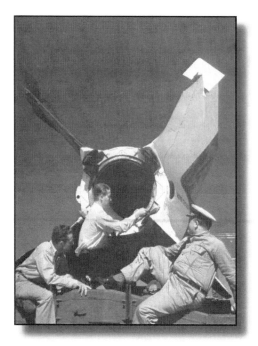

Fig 3-23 Wernher von Braun in December 1946 with V2 rocket brought to the US.

(Credit: NASA)

The von Braun story is developed in subsequent chapters, but notice for now that Fig 3-24 illustrates that Lindbergh is later having discussions with von Braun. Lindbergh had earlier written in his 1952 (i.e. 5 years *before* the first satellite had even got into space) preface to his Pulitzer Prize winning book "The Spirit of St Louis" that "we speculate on traveling through space as we once discussed flying across oceans". His thinking was certainly in the vanguard. By this time, after the end of the Second World War, von Braun's rockets were getting the US into space. During the war, von Braun had developed the liquid fuel technology, which Goddard had been pursuing experimentally, to the point of achieving practical payload-carrying vehicles.

The next series of photographs (refer to Fig 3-25, Fig 3-26, Fig 3-27) gives a further perspective on Lindbergh's continuing influence on space developments. When Lindbergh came to San Diego to commission the Ryan Airlines Corporation to make his aircraft for the 1927 Trans-Atlantic record bid, there was no such

Fig 3-24 Charles Lindbergh meeting with Wernher von Braun in November 1969.

(Credit: Lindbergh Picture Collection, Manuscripts and Archives, Yale University Library).

aircraft as the "Spirit of St Louis" on the production line. However, they did have an aircraft called the Ryan B1 Brouhgham, which the young aviator asked them to modify to make his flight possible. The main thing that he needed was more and bigger tanks for his aviation fuel. He redesigned the aircraft so that he could carry sufficient fuel for the trip. The fuel was carried straight in front of the pilot, near the center of gravity, in order for there to be minimal impact as the fuel was consumed during the flight, and to give him some chance of survival in the event of a crash.

There was, however, a major downside to this plan. It meant that Lindbergh could not see forward from his cockpit. Since most of the flight was to be over featureless ocean, Lindbergh reasoned that this tradeoff was something that he could accept for the majority of the flight. He did not, after all, expect to meet anyone coming the other way. Plus, it was commonplace at the time for pilots of "taildraggers" to have to sway from side to side during takeoff to see ahead. However, there were two parts of his trip where seeing forward would be crucial to his safety and success: low-flying and landing. So, the Lone Eagle flew with the solution that is visible in Fig 3-25, which shows the cockpit of the "Spirit of St Louis". A periscope was installed. You can see it in the image, at his eyelevel. The device stuck out of the port side of the fuselage, just in front of his side window. He looked ahead into the hole in the instrument panel, just left of center, and was thus able to see around his fuel tank and engine. We shall soon see how this simple solution received a second opportunity to help out in the pioneering of flying technologies that were constrained by weight and space considerations.

So, we continue our brief diversion into the early US space program (trust me, we will get back to aviators!). Project Mercury was the first step for the US in pursuing the dream of manned spaceflight and the Mercury capsule was just big enough to carry one person. The launch vehicle for the first flights, the Redstone, would be designed by von Braun and his team of "steely eyed missile men", as a modification of his V-2 rocket.

Fig 3-25 The cockpit of the "Spirit of St Louis", showing the periscope that Charles Lindbergh used to be able to see forwards.

(Credit: National Air and Space Museum)

Payload mass was at a premium, and the Mercury capsules therefore were very cramped. Mercury astronaut John Glenn famously said: "You do not get into a Mercury capsule: you put it on!" This was to be the first step into space and there were a great many unknowns. How would the astronaut react? How well would the newly designed spacesuit operate? Would the structure of the capsule stand up to the vibrations and pressures of the flight? Because of this last concern, it was decided that a full window for the astronaut might represent a structural weak point, and so for the first two US spaceflights, the Lindbergh periscope solution was used again. Alan Shepard was the pioneer who was the first US astronaut to fly into space, in the Mercury capsule Freedom 7. Fig 3-26 displays his cramped environment and small circular side window, which enabled light to enter the cabin but was otherwise of little value in letting the astronaut get his bearings. In front of Shepard, however, was a periscope just like that in the "Spirit of St Louis", which enabled him to see forwards. It had gradations on the cockpit screen to assist in navigation and alignment. When Alan Shepard said: "What a beautiful view!", he was enjoying the panorama provided by the periscope. Fig 3-27 introduces the second American in space, Gus Grissom, looking back down through the periscope from the outside at his fellow Mercury astronaut in the capsule. Gus Grissom did not have a great deal to say, but his admonition to the technicians who were assembling his launch vehicle and Mercury spacecraft to "Do good work!" is just as valid today as when he originally said it. Particularly in the early years of space tourism, the industry will need to be extra-vigilant on matters of safety. Early accidents will not be helpful to the nurturing that the space tourism business needs.

Fig 3-26 Inside the cramped Freedom 7 Mercury Capsule in 1961 with no forward view, except via the periscope. Alan Shepard is ready for the first sub-orbital lob into space (to become the basis for the future sub-orbital space tourism industry).

(Credit: NASA)

Fig 3-27 Mercury Astronaut Gus Grissom looks back into the Mercury capsule at his buddy Al Shepard inside, via its periscope, a Charles Lindbergh legacy. The capsule is now on display at the US Naval Academy at Annapolis, Maryland.

(Credit: NASA)

Both Alan Shepard, and subsequently Gus Grissom, flew sub-orbital trajectories not unlike those of today's sub-orbital space tourists. They were the brave pioneers who checked it out, back in 1961. And they used a Lindbergh idea in doing so. After the first test flights had proven the structural integrity of the Mercury capsule, the later missions included a fuller window in front of the astronauts. Shepard subsequently became the only Mercury astronaut to reach the Moon. He was the Commander of Apollo 14, which flew to the Fra Mauro region in February 1971 (Fig 3-28). He named his Moon ship "Kitty Hawk", in homage to the first flights of the Wright Brothers from those North Carolina sand hills in 1903. On stepping down to the lunar surface, ten years after his, and America's, first spaceflight, Shepard said: "It's been a long way, but…we're here!"

Since space tourists have now followed Shepard into space, and even into orbit, we can ask how long will it take to follow him onto the Moon? Some Moon missions for space tourists are in fact already being planned by the Space Adventures organization, about which we will learn later. And you can be sure that the spacecraft being designed for space tourism purposes certainly have plenty of window area, and don't need periscopes!

Fig 3-28 America's first man in space eventually walks on the Moon. Alan Shepard during the Apollo 14 Mission. Shepard led the way for sub-orbital space tourists, although he surely would not have realized it at the time.

(Credit: NASA)

Now at last, after our Lindbergh-induced detour into rocketry, we can get back to aviation. In 1933, the Boeing 247 was introduced as the first truly modern airliner. It was all metal, and streamlined, and had controllable pitch propellers, but it still used to vibrate in ways that would not be acceptable today! It nevertheless provided a new level of comfort for its ten passengers. Following the Second World War (1939-1945), during which again there were significant advances in aviation, surplus aircraft supplies became available that enabled new airlines to begin operations. In particular, the DC3 Dakota, otherwise known as the C-47 troop carrier, was a favorite for startup airlines (Fig 3-29). This was Douglas's 1935 answer to the Boeing 247, was faster at 190 mph and could seat 21 instead of only 10. The author had his own first flight in a Manx Airlines DC3 around 1953, while going with his parents on holiday in the UK from Newcastle to the Isle of Man, a small island in the Irish Sea, which had previously required an uncomfortable and long ferry trip. These airliners which opened up flying for all were not pressurized; which meant that they flew *through* the weather, low, slow and noisy. The author remembers that the stewardess issued hard candy to passengers to suck - to counteract the headache and earache caused by the change in pressure during take-off and landing. It was still such a new experience that everybody was photographed on the aircraft passenger steps on leaving the plane. The coming of the DC3 was an important start towards the objective of making it routine for anyone and everyone to be able to fly. Anywhere. In 1956, a DC3 fitted with skis was the first aircraft to land at the South Pole.

Fig 3-29 Douglas DC3 Dakota was used for post WW2 airline fleets, so that the general public began to be able to take to the air.

(Credit: Knowledgerush.com)

Frank Sinatra caught the mood of the era of air travel for all with his 1957 anthem to passenger flight "Come Fly with Me", with an invitation to board a Lockheed Constellation (Fig 3-30), written by Sammy Kahn with music by Jimmy Van Heusen:

" Just say the words
And we'll beat the birds
Down to Acapulco Bay.."

Fig 3-30 Frank Sinatra's invitation to fly just about anywhere via Constellation in 1957.

(Credit: Maxalbums.com)

Amongst the technological developments brought about by WW2 was the almost simultaneous invention of the jet engine in Britain and Germany. The inventors were Sir Frank Whittle and Fritz von Ohain (both incidentally in their early twenties at the time). Germany produced the Messerschmitt 262, and the British produced the Gloster Meteor, both of which were operational fighter jets which saw combat during the closing stages of the war. The author was fortunate as a cadet in the RAF Air Training Corps (ATC) to be able to fly a 600mph Meteor T7 trainer (Fig 3-31) in 1962. Going faster and higher was the name of the game, and the jet engine opened up the possibilities. The author's own lifetime interest in aeronautics and space travel was certainly fueled by that early opportunity to fly in an RAF jet, doing aerobatics in the brilliant sunlight high above an overcast rainy day in Lincolnshire.

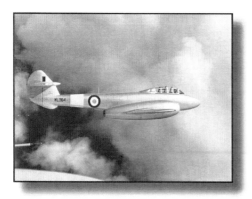

Fig 3-31 Gloster Meteor, the first British jet fighter, saw service during World War II. The jet engine was developed simultaneously in Germany and Britain during that conflict, and led to unprecedented increases in performance, both in speed and altitude.

(Credit: HistoryofWar.org)

After the war, the first jet airliners emerged, with the British de Havilland Comet leading the way in 1949. The introduction of the Comet produced a great leap forward in air travel. At around 500mph, it flew at almost 200mph faster than the best of the piston-engined airliners, such as the Douglas DC-7 and the Lockheed Constellation, whose useful operational lives were thereby curtailed. As if in recognition of the close of an era, Orville Wright had briefly piloted a Constellation in 1944, just 4 years before his death – what amazing changes he had seen! (Fig 3-32).

Fig 3-32 Orville Wright in 1944 at the controls of a Lockheed Constellation. Jetliners would subsequently take over from propliners.

(Credit: mcguinessonline.com)

The flying public could now benefit from the new jet engine technology. The author, for instance, flew by British European Airways (BEA) from London Heathrow to Paris Le Bourget via Comet (Fig 3-33). It was a big deal, with a new level of speed comfort and quiet. No sooner had we taken off than we were landing. Of course, the passenger jets would show their greatest benefits on long-distance flights. Fig 3-34 shows the first US jetliner introduced in 1958 - the highly successful Boeing 707 (of which around 800 were produced). The 707 benefited from lessons learned in early Comet accidents. It does not always pay to be first. The French introduced the Caravelle around the same period. Because these new aircraft types were not only jet-powered, but pressurized, they brought long distance flying with speed, comfort and quietness to the flying public. They could fly way *above* the weather, as Lindbergh had foreseen. The term "the jet set" entered the aeronautical cultural lexicon.

Fig 3-33 The de Havilland Comet, the world's first jet airliner.

(Credit: RAFMuseum.org.uk)

Fig 3-34 Boeing 707, the first, and highly successful, US jetliner.

(Credit: Aerospaceweb.org)

Lindbergh's business partner, Juan Trippe, was again at the forefront of aviation developments with his PanAm fleet of Boeing 707s. Pan Am's first passenger jet flight was operated in October 1958 from Idlewild (now New York JFK) to Paris. The new jets allowed PanAm to introduce lower fares and fly more passengers. The 707 flew twice as fast, and carried twice as many passengers, as the propeller driven airliner, the Stratocruiser, which it replaced. Trippe hired Lindbergh to help design the cabins, and to develop the trans-polar routes.

Fig 3-35 represents perhaps the ultimate (until today) in getting the public into the air in large numbers – the Boeing 747 Jumbo Jet. This was Trippe's biggest gamble. He had specified his needs to Boeing and guaranteed that PanAm would purchase a large fleet. He asked for an aircraft that would be two-and-a-half times the size of the 707, with 30% lower operating costs per seat. It first flew in 1969. Juan Trippe, who did so much, with Lindbergh, in creating the airline business that got the average Joe into the air, died in 1981, and his airline died a decade later of debt-induced financial crises, in 1991.

Fig 3-35 The Boeing 747 Jumbo Jet, created to serve a need identified by PanAm's Juan Trippe, brought the flying experience to large numbers of the general public, over 400 at a time.

Lindbergh continued to demonstrate his interest in the developing space program, was called upon by successive presidents for advice,

and lived long enough to witness the success of the Apollo program, watching all six of the Moon landings and meeting the astronauts who flew the missions. Fig 3-36 records one such meeting in 1968 in the White House with Lindbergh, President Johnson, and the crews of Apollo 7 (Cunningham, Eisele, Schirra) and Apollo 8 (Anders, Lovell and Borman). Nearly forty years later, Bill Anders of Apollo 8 would recall that meeting when he gave the prestigious Charles Lindbergh Memorial Lecture at the Smithsonian in Washington, DC. He said how moving it had been to have Lindbergh's support before their first ever launch away from Earth to the vicinity of the Moon. In his audience were Lindbergh's grandchildren.

Fig 3-36 White House Meeting of Charles Lindbergh (signing, right) with Apollo 7 and Apollo 8 crews (also in the photo are the President and Mrs. Johnson, NASA Administrator James Webb, and Vice President Hubert Humphrey). Dec 9th 1968.

(Credit: Lindbergh Picture Collection, Manuscripts and Archives, Yale University Library).

Fig 3-37 The Anglo-French Concorde. Flights began in 1969 and ended in 2003. Supersonic travel became possible for the (wealthy) general public.

(Credit: Airliners.net)

By July 1969 the first Moon landing had taken place. In that same year, both the 747 and the European Concorde (Fig 3-37) had flown and began operations. Passengers could now fly at supersonic speeds. Surely, it would not now be long before the public would be able to fly into space?

The last flight to the Moon was in December 1972. Charles Lindbergh had been in the mainstream of aviation and space developments since the very beginning: he had shared experiences with both the first aviator Orville Wright, and the first man on the Moon, Neil Armstrong.

But even so, his story is not yet over. His grandson Erik Lindbergh has continued Charles Lindbergh's quest. Erik Lindbergh is a private pilot, and he decided to celebrate his grandfather's achievement by himself flying solo across the Atlantic in April 2002 to celebrate the 75th anniversary of the original pioneer flight. Of course he had a modern aircraft with modern navigation aids, a Lancair Columbia 300, but it was still a memorable re-enactment of the original 1927 feat.

But this is not the main reason that the Lindbergh name continues to be associated with flying and indeed spaceflight. Erik Lindbergh was a co-founder of the Ansari X-Prize, which was a $10M prize established to encourage the creation of a space tourism industry. We shall learn more about it later. The prize was won in 2004 by Burt Rutan with his SpaceShipOne rocket plane, about which we shall also learn more later.

Fig 3-38 captures Erik Lindbergh with Burt Rutan, the aircraft designer and eventual prizewinner, at the Ansari X-Prize event in the Mojave Desert in California. Lindbergh carries a small rocket model. He is an artist, and some of his artistic output includes model rockets.

The Ansari X-Prize was modeled after the Orteig Prize which had attracted the young "Lone Eagle", Charles Lindbergh, to try his risky Trans-Atlantic flight in 1927. It ultimately led, as we have seen, to the creation of an airline industry, with flights for all.

Fig 3-38 Burt Rutan, designer of SpaceShipOne, with Erik Lindbergh at Ansari X-Prize in 2004. Charles Lindbergh's legacy motivates a new generation of adventurers in the air and space.

(Credit: Xprize.com)

Fig 3-39 Sir Richard Branson, airline chief and adventurer with fellow adventurer and record holder the late Steve Fossett.

(Credit: Jeff Cooper/ Salina Journal)

One of the industrialists who later established an airline (Virgin Atlantic Airways) to continue Juan Trippe's vision of bringing flying to an increasing proportion of the general public was Sir Richard Branson. His operation began in 1984. Branson was an adventurer and he undertook many risky flights, sometimes by balloon, with his friend the late Steve Fossett, who held many world flying records (Fig 3-39).

Fig 3-40 Round-the-World Voyager craft, built by Scaled Composites, and flown by its pilots Dick Rutan and Jeana Yeager in 1986. The flight took 9 days. The aircraft is now in the National Air and Space Museum. The technology enabled the new generation of space tourism craft such as the White Knight mothership for SpaceShipOne.

(Credit: Xpda.com/BobWebster).

Sir Richard Branson sometimes used the dry desert environment of Mojave airport to store and service part of his airline fleet, and this was the base for Burt Rutan's company Scaled Composites. Rutan was a creative and visionary aircraft designer and engineer, who was developing entirely new manufacturing technologies to make aircraft lighter and stronger. Furthermore, to demonstrate his new technologies, Rutan had built the record-breaking 1986 Voyager (Fig 3-40), and Fossett's 2005 Global Flyer (Fig 3-41) there. Because of the development of these increased efficiency aircraft manufacturing techniques and the revolutionary designs, the 1986 Rutan/Yeager crew of the Voyager could fly around the world on the contents of one (rather large!) gas tank, even though it took them 9 days to do it. Dick Rutan was, with his fellow pilot Jeana Yeager, flying the aircraft that had been built by his brother Burt. Incidentally, Jeana Yeager is not related to Chuck Yeager. Steve Fossett could do it on his own in 2005, because the jet engine power plant, and better instrumentation, navigation and control technologies, made the trip possible in less than 3 days. We shall see later just how important was Mojave, and the meeting of Burt Rutan with

Sir Richard Branson, to the possibility of space tourism for the general public.

Fig 3-41 Steve Fossett's jet-powered Global Flyer, built by Scaled Composites, returns from first solo non-stop around the world flight, March 4, 2005, 67 hours in the air.

(Credit: News.Sky.Com)

To build Voyager and Global Flyer, Rutan had developed new prototyping and manufacturing techniques, which led to major advances in strength to weight ratios, crucial for later space tourism developments. Rutan's Voyager now hangs from the ceiling at the entrance of the National Air and Space Museum at the National Mall in Washington DC, and his Global Flyer is at the NASM's Udvar Hazy Center facility in Dulles, VA. We shall see in Chapter 8 how the design of the White Knight mothercraft, which carried the prototype space tourism vehicle SpaceShipOne to its high altitude launch point, was much influenced by the technologies developed for the Voyager.

We have followed the developments in aviation from the very first flight and, often stimulated by prizes and courageous record-breaking feats, to the arrival of passenger flying for everyone. Jet technology made it possible to go ever faster and higher. We have watched as the airplane and engine technology has progressed to new materials and manufacturing techniques that bring us right to the doorstep of regular space tourism operations, and finally, the Ansari X-Prize has led to the imminent birth of a space tourism industry with space flights for all. The Lindbergh name clearly embodies the "Wright Stuff", and links the whole range of achievements and exploits in aviation and space from the very first to the most recent. In addition to the Wright

Brothers themselves, whose dedication, vision, persistence and risk-taking defines the very characteristics of the award, Charles Lindbergh gets the "Wright Stuff" award for Aviators. In his book "The Spirit of St Louis" (Ref 2) Lindbergh said: "I don't believe in taking foolish chances; but nothing can be accomplished without taking any chance at all."

You are standing on the shoulders of some courageous and creative people who made sure that the aviation part of the space tourism equation was fully understood, demonstrated, and ready for its role in space tourism. But that would not be enough in itself to get you into space.

The Wright Brothers

Charles Lindbergh

"Man is going to explore the universe, and pioneer, and settle the universe.
There's no question; it's just when.
You don't have to do it this year or next year,
but it'll get done."

David R. Scott, Commander, Apollo 15.
(from the book in Ref 8)

Now we move to the rocket men. Who were they, and what is their contribution to the creation of a space tourism industry and to your ticket to space? The rocket men were dreamers, scientists and engineers. Space, by definition, is a global enterprise, and nowadays countries all over the world are involved in some way with space developments. The author, for example, has negotiated space-related business in some forty countries. But at the beginning, the pioneers came from just a few of them. We shall largely focus on Russia, Romania, Germany, the US and UK. This is where they worked to get mankind into space. The initial spacemen who went up on rockets starting in 1961 were called astronauts in the USA and cosmonauts in Russia, and they were all at the outset military test pilots, a category which at that time excluded women. Subsequently, women also flew into space, and the qualifications moved away from military experience, embracing various scientific disciplines. We extend the definition of "rocket men" beyond the scientists and engineers to also include some of these early astronauts and cosmonauts in this chapter, although as by now you will have realized, they will not be *limited* to this chapter. They are likely to also pop up elsewhere in the book. They took risks, which, although they probably did not know it, eventually made space tourism a possibility for all. But nevertheless, they remained government employees until the birth of space tourism three decades later with the flight of the Japanese journalist Toyohiro Akiyama on board a Soyuz vehicle in 1990.

The story of aviation began, as we have seen, in 1903. Rather surprisingly, in that very same year an almost deaf schoolteacher in Kaluga, in the frozen north eastern part of Russia, wrote an important theoretical paper: "Investigation of World Spaces by Reactive Vehicles", in which he developed what are now simply called the rocket equations. His name was Konstantin Eduardovich Tsiolkovsky (Fig 4-1 and Fig 4-2), and he is generally regarded as the father of modern rocketry (see Ref 1).

Fig 4-1 Konstantin Tsiolkovsky, "Father of Modern Rocketry".

(Credit: Kaluga Museum of Cosmonautics)

You have already seen how the aviation industry developed towards making your spaceflight possible, and how a succession of important steps was needed. Now for the rocketry perspective. In order for you to fly on your space tourism trip, you will discover that a number of discreet steps in rocketry were also needed. This chapter will show that we have needed to

Name Konstantin Eduardovich Tsiolkovsky

(Credit: Kaluga Museum of Cosmonautics)

Summary Description
Pioneer rocket scientist

Date of Birth
September 17 1857, Izhevskoye, Russia

Date of Death
September 19 1935, Kaluga, Russia

Nationality Russian

Achievements
Derived the rocket equations

Specific help for Your Ticket to Space
Working independently from Goddard, and completely on his own, he developed the principles of spaceflight round about the time that the Wright Brothers were learning the principles of flight in the air. His deafness was not the result of testing rocket engines at full thrust in close proximity, but was a consequence of heredity. He was a theoretician, not a test engineer. Unlike Goddard, however, we know that Tsiolkovsky would *not* have been surprised to witness spaceflight. He was pretty certain that it would happen, and that the human race needed it for its long term survival. Sputnik 1 flew into orbit 22 years after his death, and it only took 34 years after his death to put a man on the Moon. And his writings were directly responsible for this outcome. The causation path goes directly from Tsiolkovsky via Oberth to Korolev and von Braun. No doubt about it, your ticket to space owes a lot to this man's vision and clear analysis, in a place and time when for most of us it would be too cold to think!

advance from designing and building rockets and spacecraft, to developing commercial space business, to flying living beings, to crewed spacecraft, to maneuvering, rendezvous and docking in orbit, to space station assembly in orbit, to re-usable crewed craft and thence to giving rides to paying customers. No step could have been left out if we were going to have our own ticket to space. Let's follow this journey with the Rocket Men.

Fig 4-2 Tsiolkovsky 10 Kopek postage stamp.

(Credit: Stamprussia.com)

From the 1903 initial paper, Tsiolkovsky continued to refine his work, producing papers until his death in 1935. In his 1912 update of his original paper, he reports (Ref 1) that: "….the first seeds of the idea were cast by the famous fantasy writer Jules Verne; he awakened my mind in this direction…."(Author note: Verne's "From the Earth to the Moon" had been published in 1865), and: "…the conquest of the air will be followed by the conquest of ethereal space….", and: "….the better part of humanity will never perish but will move from sun to sun as each one dies out in succession….." So Tsiolkovsky envisioned from the start the possibility of spaceflight for all. His vision even took him *beyond* space tourism, to embrace the notion of eventual mass migration to space. He was aware that at some distant point in the future, mankind would not be able to continue to live on this planet (due to changes in the Sun, bombardment from asteroids, etc.). Fig 4-3 displays his Moscow monument, just outside that city's famous astronautics museum.

Fig 4-3 Monument to Tsiolkovsky in Moscow.
(Credit: Graham Chand)

Recall from the previous chapter that Robert Goddard in the US published his own seminal paper "A Method of Reaching Extreme Altitudes" in 1920 (see Fig 4-4 for Goddard retrieving rocket nosecone). Goddard's vision was perhaps more prosaic than Tsiolkovsky's, although he knew it would be possible to get a rocket to the Moon. He had even experimented on how much flash powder it would need to carry so that the impact might be seen from Earth! Both men, in their very different environments, set in motion the practical steps needed to get mankind into space.

Fig 4-4 Robert Goddard retrieving nose cone from rocket test flight, Roswell, New Mexico, 1937.

(Credit: NASA)

When the former Soviet Union launched the first spacecraft to see the far side of the Moon, Lunik III in October 1959, the image sent back to Earth, notwithstanding the magnitude of the engineering feat, was nevertheless rather blurred, and lacking in detail. However, one thing was very clear: there was a very significant feature on the far side, which happened to be very large and was distinctively colored. It was certainly the most prominent lunar far side feature that could be discerned from those first primitive images, and so the Soviets named this feature for Tsiolkovsky. Subsequent flights, including those from the American Apollo missions, confirmed the nature of the feature, which turns out to be a lunar *mare* surrounded by mountainous terrain, and with much darker coloration than its surroundings. Fig 4-5 reveals Tsiolkovsky Crater photographed from an Apollo spacecraft, giving tangible, permanent and appropriate recognition to the father of modern rocketry for his achievements. Goddard also has a crater on the moon named in his honor, although it is not nearly as significant a landmark!

Fig 4-5 Crater Tsiolkovsky, Lunar far side, photographed from an Apollo spacecraft.

(Credit: NASA)

We shall see that Tsiolkovsky's influence was enormous. He certainly had a lot to do with the initial lead that the Soviets had in space in the 1950's and 1960's. Sergei Pavlovich Korolev (Fig 4-6) was the rocket engineer who later emerged out of the secrecy of the Soviet Union, known originally only as the Chief Designer. He was responsible for the Vostok rocket that launched both the first artificial Earth satellite Sputnik 1, in October 1957, and also the first man in space Yuri Gagarin, in April 1961 (see Ref 2, Ref 3) in a Vostok spacecraft (Fig 4-7). Behind Korolev in Fig 4-6 is Sputnik 1, which caused so much alarm in the USA when it was launched into orbit. The satellite carried very little instrumentation, but did transmit a "…beep beep beep…" signal as it flew around the world, making its presence very obvious.

Fig 4-6 Sergei Korolev, Russian chief rocket designer.

(Credit: Wikimedia)

Fig 4-7 Vostok capsule for first man in space Yuri Gagarin, designed by S.P. Korolev.

(Credit: Energia)

In September 1957, just a month before the launch of the first satellite Sputnik 1, and thereby the start of what has become known as the Space Age, Korolev, then unknown in the West, gave a lecture. It was given on the Centennial of Tsiolkovsky's birth, with the title: "On the Practical Significance of the Scientific and Engineering Propositions of Tsiolkovsky in Rocketry" (Ref 1). Korolev made it clear how important was Tsiolkovsky's work, and one can only speculate on how many of the audience in that secretive time and place knew what was about to happen, and how it would transform the world.

After the worldwide impact of Sputnik 1, Korolev then built up experience of the effect of spaceflight on living beings by flying a succession of dogs on more and more ambitious missions. Laika flew in Sputnik 2 just a month after Sputnik 1, and survived the experience although the capsule was not recoverable. Later, in March 1961, the dog Zvezdochka (Starlet) was successfully recovered from orbit, and appeared to have suffered no ill effects from the experience. So the next step would be to find out if humans could also survive in space.

Fig 4-8 shows Korolev later with Yuri Gagarin, after the latter's pioneering 12th April 1961 earth-orbit spaceflight in Vostok 1. Gagarin was only 27 when he undertook the flight.

Fig 4-8 Yuri Gagarin, first man in space, with S.P. Korolev, designer of his rocket and spacecraft,1961.

(Credit: Energia)

Fig 4-9 records Gagarin inside his Vostok, and Figs 4-10 and 4-11 provide some evidence of the fame resulting from the first manned space flight. Can you imagine the courage it took to sit in an enclosed sphere, on top of a rocket filled with tons of highly inflammable fuel and oxidizer, all on your own, to be the first man in space? Gagarin came to London on his world tour, and received a medal from the British Interplanetary Society (BIS), and there were enormous crowds wherever he traveled. Although Gagarin himself did not know it, an essential next step towards space tourism had been achieved. Man could survive in space. Unfortunately, Gagarin died only 7 years after his pioneering Vostok 1 flight, while taking part in a routine training flight in a Mig 15 fighter aircraft.

Decades later, new generations re-connected with the events of April 1961 when Loretta Hidalgo and George Whitesides created "Yuri's Night", an annual youth-oriented celebration of Gagarin's bravery and achievement. Whitesides subsequently headed up the National Space Society (which had been created by von Braun) and its annual jamboree, the International Space Development Conference (ISDC), where space tourism has been a regular topic for many years. In 2009 he joined NASA as Chief of Staff as part of the changes made by the Obama Administration. Hidalgo and Whitesides were among the very first to sign up for seats when the formation of Virgin Galactic was announced to provide space tourism opportunities, as we shall see later.

Fig 4-9 Gagarin inside his Vostok capsule – first man in space, 12th April, 1961. His courage opened up the possibility of human space exploration.

(Credit: Spaceflighthistory.com)

Name Sergei Pavlovich Korolev

(Credit:Kapyar.ru)

Summary Description
Soviet Russian rocket engineer

Date of Birth
Jan 12th 1907, Zhytomyr, Ukraine

Date of Death
Jan 14th 1966, Moscow, USSR

Nationality Russian

Achievements
First satellite in space (Sputnik 1). First living being in space (Sputnik 2, dog Laika). First man in space (Vostok 1, Gagarin). First spacewalk (Voskhod 2, Leonov).

Specific help for Your Ticket to Space
The statue captures the determination and resolve of the engineer who labored to create the Soviet space program, and achieve the string of space firsts, including first satellite and first man in space. Korolev was known as the "Chief Designer" – he was the Soviet von Braun. And like von Braun, he was a natural leader who understood not only engineering but politics too. He was only 59 when he died of a medical mistake during routine surgery. It is hard from this distance to reliably speculate on how Korolev would have viewed the arrival of space tourism. It is undeniable, however, that all orbital space tourists have used Korolev's training and launch facilities, and that he first demonstrated that man (and women!) can live in space.

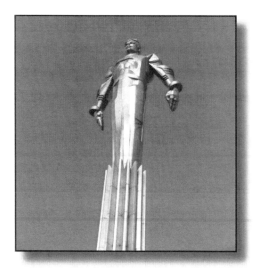

Fig 4-10 Gagarin monument in Moscow.

Fig 4-11 Yuri Gagarin postage stamp.

The Space Age of course ushered in much more than manned spaceflight. We owe our weather forecasting, much of our international communications, TV broadcasting, navigation, and more, to the technologies that resulted from the work of the Rocket Men. And Korolev's work led directly to the first space tourism flights, which used the Soyuz launch vehicle, an upgrade of his original Vostok rocket used for Gagarin's launch (Fig 4-12).

As another "first" in the space race publicity campaign, Korolev launched Valentina Tereshkova into space on 16th June 1963 in Vostok 6 (Fig 4-13). Tereshkova, as the first woman in space, would also receive a gold medal from the BIS (in 1964). The author met her, and also fellow cosmonaut and first space-walker Alexei Leonov, in New Jersey in 2005 at a space convention. Both of the Russians went out of their way to greet the formerly competitive US

astronauts Schirra and Aldrin, who were also at the convention. Leonov, as head of cosmonaut training in Russia, had taken a special interest in the early space tourism flights of the western millionaires who came to use the Russian training facilities in the early nineties. Tereshkova, through her interpreter, discussed the rigors of training and the space suit systems she had used on her mission. Leonov did not need an interpreter. He had learned English (or Oklahoman, he would say) when he worked with Tom Stafford during the joint US/USSR Apollo Soyuz Test Project in 1975 (see later in Fig 4-44). Stafford and Leonov remain firm friends to this day.

Fig 4-12 Gagarin is launched on a Vostok rocket, April 12th, 1961.

It would be 20 years after Tereshkova's courageous solo flight before the US launched a female astronaut, Sally Ride, as one of the members of the crew aboard a Space Shuttle. Nowadays, of course, we have female Space Shuttle commanders and female commanders of the International Space Station. The very idea of recording that a woman "could" be an astronaut seems almost outrageous, and certainly condescending. However, we must remember the social mores of the time, and the work of the pioneers like Valentina Tereshkova; who had to deal, not only with the rigors of space travel, but also with male prejudice. Amongst the aviators, brave pioneers like the Americans Amelia Earhart and Anne Morrow Lindbergh, and the British Amy Johnson had had to face the same kind of problems in the Twenties and Thirties. Aviation and space travel have been of some help in changing the old stereotypes about female roles, but only because those early female pilots and astronauts stood up to the mark, and took the risks alongside their male counterparts. Space tourism is an equal-opportunity endeavor.

Fig 4-13 Yuri Gagarin and Valentina Tereshkova, space pioneers, after her June 1963 flight, the last one of the one-person Vostok spacecraft.

(Credit: Rusmedia/Eurorus4en.wordpress.com)

We have just seen something of the Soviet rocket program. Now we turn to look at the German program. We have already mentioned Wernher von Braun. He was the German designer of the V2 rocket that was used against the British during the Second World War. We shall learn more of him later, but suffice to note for the present that he followed the lead of his mentor Hermann Oberth (Fig 4-14 and Fig 4-15). Oberth had in turn studied the work of Tsiolkovsky. Oberth was a Romanian/German scientist, medical doctor, visionary and rocket experimenter. Apparently, he was totally unaware of the contemporary work of the somewhat reclusive Goddard in the USA at that time. Towards the end of the Second World War, London was on the receiving end of von Braun's rockets. Members of the British Interplanetary Society, which was based in London, could see their effectiveness first hand. The Society had been founded in 1933 in Liverpool, but had moved to London in 1937 (see Ref 4). The first rocket society had been formed in Russia in 1924, and Tsiolkovsky was a member, and in 1931 Korolev joined its successor organization. In 1927 in Germany the *Verein fur Raumschiffahrt* (VfR), i.e. the Society for Space Travel, was formed. Its membership included Oberth and von Braun (see Ref 5, Ref 10).

Name Hermann Julius Oberth

(Credit:AIAA.org)

Summary Description
Pioneer Rocket Engineer

Date of Birth
25 June 1894, Sibiu, Romania

Date of Death
28 December 1989, Nurenberg, Germany

Nationality Romanian/German

Achievements
Concept of multi-stage rocket.
Wrote *"Die Racket zu den Planetenraum"*, 1923

Specific help for Your Ticket to Space

Hermann Oberth was the mentor of Wernher von Braun, and you know what *he* was able to achieve. Oberth's famous 1923 book captured the imagination of people all over the world. Oberth's ideas led to the creation of the *Verein fur Raumschiffahrt* (VfR), the Society for Spaceship Travel, with its rocket flying field in Reinickendorf, Berlin, for its experiments. The press was invited to the tests, increasing public awareness. Von Braun learned his craft with Oberth, and then went on to design the V2 during the Second World War, the first really practical space rocket. So, your ticket to space would not be worth very much if Hermann Oberth had not worked to convert Tsiolkovsky's ideas into engineering reality. Going by the public records, Oberth was a serious man, never known to smile throughout his 95 years. We can only hope and believe that might surely have changed if he had his own ticket to space!

Fig 4-14 Hermann Oberth, rocket pioneer, and subsequent director of motor development at Peenemünde.

(Credit: Meaus.com)

Fig 4-15 Oberth commemorative postage stamp.
(Credit: Liceodavincimaglie.it)

The American Interplanetary Society, subsequently named the American Rocket Society, was formed in 1931. So there was something of an international *camaraderie* of rocket men (and women) in the early years, with correspondence between the various national societies. A member and subsequently the President of the British Interplanetary Society during the time that up to 3,000 V2s were raining down on London, was one Arthur Charles Clarke (Fig 4-16).

Fig 4-16 Sir Arthur C Clarke at his desk, after moving to his adopted country, Sri Lanka, to live.
(Credit: Amy Marash)

Clarke was a visionary, a futurist and author. He had a major influence on the development and commercial use of space technologies, as we shall see. Fig 4-17 exhibits an article he wrote to Wireless World in 1945, while the Second World War was still in progress, suggesting using the surplus military V2 rockets for space research in the post-war era. Imagine that! They were raining down on his city London. What a profoundly different way visionaries have of looking at the world around them.

Not satisfied with just one breakthrough concept, Clarke soon after wrote to the same publication describing the idea of satellite communications and the geo-stationary orbit. The orbit at 36,000 miles above the equator where an orbiting spacecraft takes precisely 24 hours to circle the Earth, and thus appears to be motionless to an observer on the ground (Fig 4-18). Later on it would be named after him.

So this is how the next important step leading to space tourism, i.e. the introduction of the *commercialization* of space, could begin. The first geo-stationary communications satellite was Early Bird in 1965, twenty years after A.C. Clarke's article. Nowadays, telecommunications satellites occupy the entire geo-stationary orbit. Some of them are governmental spacecraft; most of them are commercial. The author, himself a Fellow of the BIS, owes his former president A.C. Clarke for a career in satellite telecommunications that would not have been possible without Clarke's creative vision.

58   **Wireless World**   FEBRUARY, 1945

Letters to the Editor

Peacetime Uses for V2 · FM Protection Against High-Amplitude Interference Pulses · Bad Books

V2 for Ionosphere Research?

ONE of the most important branches of radio physics is ionospheric research and until now all our knowledge of conditions in the ionosphere has been deduced from transmission and echo experiments. One of the more modest claims of the British Interplanetary Society was that rockets could be used for very high altitude investigations and it will not have escaped your readers' notice that the German long-range rocket projectile known as V2 passes through the E layer on its way from the Continent. If it were fired vertically without westward deviation it could reach the F_1 and probably the F_2 layer.

The implications of this are obvious: we can now send instruments of all kinds into the ionosphere and by transmitting their readings back to ground stations obtain information which could not possibly be learned in any other way. Since the weight of instruments would only be a few pounds—as compared with V2's payload of 2,000 pounds—the rocket required would be quite a small one. Its probable take-off weight would be one or two tons, most of this being relatively cheap alcohol and liquid oxygen. A parachute device (besides being appreciated by the public!) would enable the rocket to be re-used.

This is an immediate post-war research project, but an even more interesting one lies a little farther ahead. A rocket which can reach a speed of 8 km/sec parallel to the earth's surface would continue to circle it for ever in a closed orbit; it would become an "artificial satellite." V2 can only reach a third of this speed under the most favourable conditions, but if its payload consisted of a small one-ton rocket, this upper component could reach the required velocity with a payload of about 100 pounds. It would thus be possible to have a hundredweight of instruments circling the earth perpetually outside the limits of the atmosphere and broadcasting information as long as the batteries lasted. Since the rocket would be in brilliant sunlight for half the time, the operating period might be indefinitely prolonged by the use of thermocouples and photo-electric elements.

Both of these developments demand nothing new in the way of technical resources; the first and probably the second should come within the next five or ten years. However, I would like to close by mentioning a possibility of the more remote future—perhaps half a century ahead.

An "artificial satellite" at the correct distance from the earth would make one revolution every 24 hours; i.e., it would remain stationary above the same spot and would be within optical range of nearly half the earth's surface. Three repeater stations, 120 degrees apart in the correct orbit, could give television and microwave coverage to the entire planet. I'm afraid this isn't going to be of the slightest use to our post-war planners, but I think it is the *ultimate* solution to the problem.

> ARTHUR C. CLARKE,
> British Interplanetary
> Society.

Frequency Modulation

WHILE post-war plans for television and UHF sound broadcasting are under discussion, it is important that the pros and cons of FM should be understood. Space will not permit a full discussion here; but I wish to correct a misconception, found even among responsible engineers, that FM can give no protection against ignition noise or other similar pulses which have an amplitude much greater than that of the signal carrier. The actual response of an FM receiver to very powerful impulsive interference can be summarised as follows:—

(1) In the absence of a signal, the FM receiver gives no output from impulsive interference.

(2) In the presence of an unmodulated carrier to which the FM receiver is accurately tuned, the impulsive interference causes no audible output. If the receiver is not accurately tuned, there will be an audible output, but the amplitude of the pulses in the audio-frequency circuits of the receiver will correspond to a modulation of the carrier of less than 100 per cent., in fact to a modulation depth equal to the ratio of the frequency error in tuning to the frequency swing corresponding to full modulation of a frequency-modulated signal.

(3) In the presence of a frequency-modulated signal to which the receiver is accurately tuned, the audio-frequency noise pulses are limited to the *instantaneous* level of signal modulation. If the receiver is not accurately tuned, the amplitude of the audio-frequency pulses will be increased by the amount defined in (2) above.

If it is true, as sometimes suggested, that ignition noise is the chief trouble in UHF broadcasting, this summary provides a basis for the comparison of FM with other systems, such as wide-band AM with audio-frequency limiting.   D. A. BELL.
London, N.21.

"New Thoughts on Contrast Expansion"

EXPEDIENCY be damned. My condemnation of contrast expansion was not based upon noise and neighbour tolerances. John B. Rudkin (your January issue) says "condemn the Philadelphia Orchestra because it is too large to play in the village hall." The truth is that anyone who asks it to do so should be condemned, and those who try to get the B.B.C. Orchestra into their bedroom are committing a crime. If the room is small, acoustically small, then only a limited contrast is proper, and all music

Fig 4-17 Letter by A.C. Clarke to Wireless World in February 1945 introducing the idea of artificial satellites.

October 1945   **Wireless World** .   305

EXTRA-TERRESTRIAL RELAYS

Can Rocket Stations Give World-wide Radio Coverage?

By ARTHUR C. CLARKE

ALTHOUGH it is possible, by a suitable choice of frequencies and routes, to provide telephony circuits between any two points or regions of the earth for a large part of the time, long-distance communication is greatly hampered by the peculiarities of the ionosphere, and there are even occasions when it may be impossible. A true broadcast service, giving constant field strength at all times over the whole globe would be invaluable, not to say indispensable, in a world society.

Unsatisfactory though the telephony and telegraph position is, that of television is far worse, since ionospheric transmission cannot be employed at all. The service area of a television station, even on a very good site, is only about a hundred miles across. To cover a small country such as Great Britain would require a network of transmitters, connected by coaxial lines, waveguides or VHF relay links. A recent theoretical study[1] has shown that such a system would require repeaters at intervals of fifty miles or less. A system of this kind could provide television coverage, at a very considerable cost, over the whole of a small country. It would be out of the question to provide a large continent with such a service, and only the main centres of population could be included in the network.

The problem is equally serious when an attempt is made to link television services in different parts of the globe. A relay chain several thousand miles long would cost millions, and transoceanic services would still be impossible. Similar considerations apply to the provision of wide-band frequency modulation and other services, such as high-speed facsimile which are by their nature restricted to the ultra-high-frequencies.

Many may consider the solution proposed in this discussion too far-fetched to be taken very seriously. Such an attitude is unreasonable, as everything envisaged here is a logical extension of developments in the last ten years—in particular the perfection of the long-range rocket of which V2 was the prototype. While this article was being written, it was announced that the Germans were considering a similar project, which they believed possible within fifty to a hundred years.

Before proceeding further, it is necessary to discuss briefly certain fundamental laws of rocket propulsion and "astronautics." A rocket which achieved a sufficiently great speed in flight outside the earth's atmosphere would never return. This "orbital" velocity is 8 km per sec. (5 miles per sec), and a rocket which attained it would become an artificial satellite, circling the world for ever with no expenditure of power—a second moon, in fact.

the atmosphere and left to broadcast scientific information back to the earth. A little later, manned rockets will be able to make similar flights with sufficient excess power to break the orbit and return to earth.

There are an infinite number of possible stable orbits, circular and elliptical, in which a rocket would remain if the initial conditions were correct. The velocity of 8 km/sec. applies only to the closest possible orbit, one just outside the atmosphere, and the period of revolution would be about 90 minutes. As the radius of the orbit increases the velocity decreases, since gravity is diminishing and less centrifugal force is needed to balance it. Fig. 1 shows this graphically. The moon, of course, is a particular case and would lie on the curves of Fig. 1 if they were produced. The proposed German space-stations

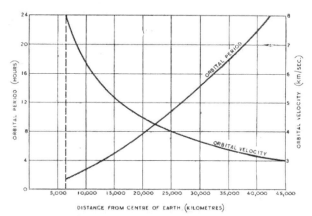

Fig. 1. Variation of orbital period and velocity with distance from the centre of the earth.

The German transatlantic rocket A10 would have reached more than half this velocity.

It will be possible in a few more years to build radio controlled rockets which can be steered into such orbits beyond the limits of

would have a period of about four and a half hours.

It will be observed that one orbit, with a radius of 42,000 km, has a period of exactly 24 hours. A body in such an orbit, if its plane coincided with that of the

Fig 4-18 Article in October 1945 Wireless World by A.C. Clarke introducing the theory of the geostationary orbit for communications satellites.

Several international organizations were formed, during the cold war era (1945 – 1991), whose purpose was the facilitation of international satellite telecommunications (including originally telex and telephony, but moving on to broadband data transfer and TV broadcasting). The United Nations was the parent organization, but the International Telecommunications Union (ITU) was set up to allocate frequencies and allocate space on the Clarke orbit for spacecraft to operate. Although created for the administration of somewhat mundane matters, such as frequency allocation, these organizations helped foster mutual respect between the representatives of the participating nations. These organizations, although established via inter-governmental treaties, nevertheless operated with a commercial mandate.

Fig 4-19 places us in a meeting in the mid-1980's of the International Mobile Satellite Organization (Inmarsat) in London, with the proceedings being translated simultaneously into 4 different languages. This was a surprising place to be during the Cold War. For the first decade of its existence, a board governed Inmarsat with representatives of 22 nations, having both Soviets and Americans amongst their members. In some respects, it can be said that via Inmarsat the Soviets learned something about capitalism and certainly about high technology procurement processes. We shall see some of the consequences later in this narrative. The author was responsible for acquiring satellites and launch vehicles for this organization. He also testified on procurement decisions, pricing, forecasts and other matters to this unusual body; which was created out of a UN concern for safety of life at sea, yet was operating in the high technology space arena with a commercial mandate.

Fig 4-20 shows one of the Inmarsat spacecraft under construction. Notice the scale. The Technical Director for the Inmarsat 2 satellite was Marty Votaw, who had been the technician soldering the parts together on the small Vanguard satellite in 1958. Marty's first satellite was only 6 inches in diameter. When the rocket had blown up on the pad, Marty had been able to retrieve the battered spacecraft from amongst the rocket debris and put it on the back of his pickup truck.

Name Sir Arthur C. Clarke

(Credit: British Interplanetary Society/www.BIS-spaceflight.com)

Summary Description
Space visionary and author

Date of Birth
16 December 1917, Minehead, UK

Date of Death
19 March 2008, Colombo, Sri Lanka

Nationality UK

Achievements
President of British Interplanetary Society
Science Fiction author
Discoverer of the Geostationary orbit

Specific help for Your Ticket to Space
Arthur C Clarke had the rare experience of being treated as a guru and futurist within his own lifetime. He brought about the commercialization of space by his discovery of the geostationary orbit, which was subsequently used by hundreds of communications satellites. Without commercial space there could be no space tourism. His 1951 book "The Exploration of Space" brought before the public the early designs of the BIS, including man-carrying rockets, orbital rendezvous and re-fuelling, lunar landing, lunar base and a Martian base. Most of this has come to pass. Clarke spread the message of space travel via a series of best-selling science fiction books, such as "2001: A Space Odyssey" which became a successful movie. If you think space tourism is still too expensive for you, you may have to wait another half century for the remainder of Clarke's ideas, such as for example the space elevator, to become engineering reality.

He found that it was still working despite the rough treatment. That satellite was subsequently donated to the Smithsonian, where it can be seen today. By the time the author worked with Marty 25 years later to define the requirements for the communications satellite Inmarsat 2, that spacecraft, with its solar arrays extended, would span 45 feet from tip to tip. Marty, like many of the early cold war missile men, found it hard to adapt to the changes brought about by the advent of commercial satellite communications. Instead of being a national priority where cost was not an issue, it had become a business operation - and an international one at that! The author had to arm-wrestle the Technical Director, who was leading the acquisition process, to even be allowed to obtain cost data as part of the procurement action. "Why do you want to know the cost of every grain of sand?...", he asked the author "...when all you want to do is buy a beach!" Of course, satellite communications became a very competitive industry, and is in fact still the *only* space-related business to become highly profitable without government subsidy.

You know very well that we expect space tourism to change all that real soon! Other international organizations operating satellites in the Clarke orbit include Intelsat, Eutelsat, Asiasat, and now many others. These organizations have made their contribution to the commercial use of space, to global communications, and to the creation of an international environment where it becomes possible to do business in space. When Marty finally retired from the scene of building spacecraft, the newspaper announcement simply noted "Marty Votaw Goes Geostationary"! Ref 9 provides a good account of the development of some of these international satellite telecommunications organizations. Although they began as quasi-governmental bodies, they have now all been privatized, and operate purely as commercial operations. New purely commercial operators have now emerged, with not even a heritage of governmental establishment.

Fig 4-19 Meeting of the governing body of the International Mobile Satellite Organization, Inmarsat, in London. Satellite communications was the first successful commercial space business. It developed during the Cold War era, and required global agreements in order to operate. It incidentally provided a market for commercial launch vehicles, such as the Delta, Atlas, Ariane and Proton vehicles.

(Credit: Inmarsat)

subsequent success of "Ariane" from Kourou in French Guiana. The Arianespace organization subsequently became the most successful provider of commercial launch services in the world.

During the author's time there, Inmarsat acquired launch vehicles from providers in the US (Fig 4-21), in Europe and Russia, and even negotiated for space launches from China. In a very real sense, cooperation in space was a forerunner of global cooperation in other fields of endeavor. And the cooperation at government level simultaneously led to global commercial space business. Commercial space, which began with satellite communications, would lead eventually to the global business of space tourism.

Fig 4-20 A typical commercial telecommunications satellite under construction. This is an Inmarsat 2 spacecraft at the factory in Stevenage, UK, March 1989. One set of solar arrays has still to be installed.

(Credit: Inmarsat)

The growth of commercial satellite communications in turn created a need for a range of commercial launch vehicles. This demand was met initially by the USA, Russia, and Europe. The USA and Russia were able to commercialize their military launch vehicles. Europe had to start virtually from scratch. The author had earlier worked on the engineering of the first European launcher, called "Europa", which formed the experience base for the later highly successful "Ariane" series. "Experience base" means, of course, that we had lots of failures to learn from, but at least each one was different. It was a tough learning curve. There were 8 flights when upper stages were added to the British "Blue Streak" to create "Europa' and attempt to launch a European satellite. We were trying to develop a launch vehicle that could carry both military and commercial payloads. To develop the European vehicle, 6 nations produced parts that all had to integrate together, and everyone spoke different languages and used a variety of units. A significant benefit from the process was simply learning how to work together in those circumstances. The lessons learned from the series of "Europa" launches from Woomera in Australia ensured the

Fig 4-21 Launch of a geostationary communications satellite at Kennedy Space Center on March 8th 1991 by one of a series of Delta 2 commercial launch vehicles purchased by Inmarsat. NASA has also used the Delta 2 vehicle to send spacecraft explorers to Mars.

(Credit: Inmarsat)

The creation of a commercial space sector necessitated the simultaneous establishment of key services essential to running any new business. Thus a space insurance industry emerged, as did commercial space attorneys and space venture capitalists, all of whom would prove to be needed for the eventual emergence of a space tourism industry.

As if to underline how the early rocket scientists and engineers valued each others' contributions, in 1949, just 4 years after London was bombed by the V2s, the British Interplanetary Society awarded both Hermann Oberth and Wernher von Braun Honorary Fellowships of their Society. Also thus awarded was Eugen Sänger (Fig 4-22), another early member of the VfR, and subsequent designer of the first concept for a point-to-point spaceplane (which would leave one spaceport, almost reach orbit, then descend at another location half a world away in less than an hour). His designs are still being considered today for "second generation" sub-orbital space tourism applications. Krafft Ehricke (Fig 4-23), another of the von Braun team, and subsequently an advisor on the design of the Space Shuttle, was honored later by the BIS (in 1974). Of course, these pioneers received awards from many other institutions and societies, too. At the end of the Second World War, the victors wanted to acquire the new rocket technology from the remains of the German rocket program. We shall read later of how the US government was able to bring von Braun, some key members of his team, and some V2's to America. Other German rocket engineers with V2 skills, such as the guidance expert Helmut Grottrüp and his team, went to the Soviet Union, and worked for Korolev. It has been said that the future "space race" between the US and the USSR would effectively be enjoined between the German engineers on each side, although the Soviets were generally more independent in their approach, using their German engineers mainly as consultants.

What did A.C. Clarke and the BIS do next? The society, which was an entirely amateur body, did groundbreaking work on the design of manned rocketships and the design of Moon landing craft. Fig 4-24, for example, provides the design of the BIS lunar lander of 1947, which bears more than a superficial resemblance to the US lunar module of 1969, twenty-two years later. We shall see further in the book that it even foreshadows the designs of some space tourism vehicles (see Ref 6).

Fig 4-22 Eugen Sänger – an early theoretician on point-to-point sub-orbital flight.

(Credit: NMSpacemuseum.org)

Fig 4-23 Kraft Ehricke describes early space shuttle concepts to TV journalist Walter Cronkite, September 1966.

(Credit: 21stcenturysciencetech.com)

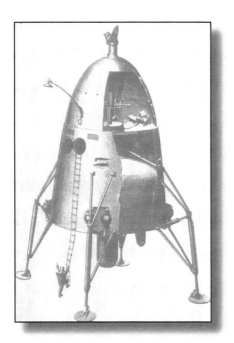

Fig 4-24 BIS Lunar Lander design of 1947 (Drawing by R.A. Smith).

Fig 4-25 Passenger shuttle leaves the Space Station. This reproduction of the original 1967 painting, has been modified by artist Bob McCall to include the Discovery and the Pod; also featured in the movie. The original painting is oil on board, and resides in the collection of the National Air and Space Museum.

A.C. Clarke was a popularizer *par excellence* of space travel, and his science fiction has spread the idea of spaceflight, and its many possible futures, to a wide general public. The 1968 MGM Kubrick movie "2001: a Space Odyssey", based upon his novel, perhaps best brought home the possibilities. Fig 4-25 reminds us of that memorable iconic image of the passenger spaceliner leaving the docking port of a giant circular space station, modeled after a von Braun design. The painting was created by Bob McCall in 1967, and was a powerful symbol of hope, a visual affirmation that humanity has the resources and the ingenuity to prosper in the 21st century. We learn more of McCall in Chapter 7.

In the movie, we see passengers arriving on board the space station from the latest shuttle flight from Earth, and it all seems a matter of routine. The space shuttle even carried the logo of a familiar airline of the time, Lindbergh-associate Juan Trippe's PanAm. Once the moviegoers had seen these images, it no longer seemed an impossibility for everyone to be able to one day go into space. During the period when Clarke was president of the BIS, of course, there had been no space flight. But he lived to see many of his predictions come to pass. For instance, satellite communications today constantly connect all parts of the Earth, and many of the spacecraft that perform the work are located in the geo-stationary orbit, which we have noted is sometimes known as the Clarke Orbit.

In Fig 4-26 we see him with Alexei Leonov, one of the early Russian cosmonauts whom we have mentioned, and a space artist, during Clarke's 90th birthday celebrations. Leonov is another of those special individuals who occupy more than one category. He was one of Gagarin's colleagues in the first cosmonaut training class (March 1960) at the start of the space program. He is also an artist who has recorded those extraordinary pioneering events, and later was in a position to help the early space tourists get trained for their flights. Arthur C. Clarke died on 19th March 2008 in Colombo, Sri Lanka, after "having completed 90 orbits of the sun", as he had said at his ninetieth birthday. Much of what he had predicted has come to pass. Some of his ideas (e.g. the space elevator for getting cargoes into orbit) are yet to see their engineering fulfillment, and of course might never happen.

Fig 4-26 Cosmonaut Alexei Leonov presents some of his space art to A.C. Clarke on Clarke's 90th birthday in Colombo, Sri Lanka, 16 December 2007, a few months before Clarke's death.

(Credit: Thilina Heenatigala)

Fig 4-27 Early BIS meeting (July 1938) with A.C. Clarke (far right) and Robert Truax (with rocket). Also in the photo, behind Truax's left shoulder, is the space artist R.A. Smith. The meeting took place in Smith's home.

(Credit: British Interplanetary Society/ www.BIS-spaceflight.com)

While we are talking about A.C. Clarke and the BIS, there is another Rocketman who deserves mention as a popularizer of space and rocket travel. Fig 4-27 records a pre-war image from July 1938 of members of the BIS, with A.C. Clarke to the right, greeting a visiting member of the American Rocket Society (see Ref 4). The visitor was Midshipman Robert Truax, and he has in his hands his 200 HP liquid fuel rocket motor. He did important early work for the US Navy in rocket development, and later, during the Second World War, developed rocket-assisted take-off for planes operating near their limits on runway length and payload. He also worked on the design of Atlas and Polaris rockets.

Fig 4-28 Meeting of Project Orbiter in March 1955. Von Braun is at front and Robert Truax is in the back row, 3rd from left.

(Credit: NASA)

We see Truax again in Fig 4-28, as a member of the team planning the first US satellite launch. He continued throughout the twentieth century to try and develop cheap commercial rockets which he intended would be used for space tourism purposes. He called this concept his Volks-Rocket. One of his team members was Jeana Yeager, who we saw in Chapter 3 went on to become a record breaking test pilot of the Voyager aircraft. The author, as a consultant, met with Truax at his facilities in San Marcos, CA, in 1998, where he was still working on pressure-fed, simple rockets intended to be reusable. He was still full of energy (he was then in his seventies). In 2002 he was still submitting papers on rocket technology to space conferences.

Perhaps Truax's best-known rocket, however, appears in Fig 4-29. Truax was the designer and builder of the rocket engine for daredevil Evel Knievel's "rocket bike" in which he attempted to launch himself over the Snake River Canyon in September 1974. This event was publicized widely and took up prime time TV coverage. The rocket worked, but an emergency parachute was deployed too early and prevented the stuntman from clearing the canyon on his "Canyon Jump" attempt. This did however make Evel Knievel the first private US citizen to ride a rocket.

Fig 4-29 Evel Kneivel (striped shirt) is first private US rocket rider, September, 1974. Truax at left built the motor.

(Credit: The Vintagent/Paul D'Orleans)

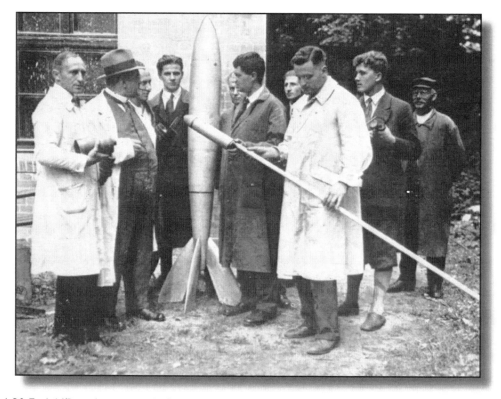

Fig 4-30 Early VfR rocket test with Oberth (center) and von Braun (2nd from right).

(Credit: V2rocket.com)

As the threads of this story intertwine, bringing us closer to space tourism, we need to step back again to look closer at von Braun's contribution. In Fig 4-30 we see the young von Braun (second from the right) preparing for a rocket test at the VfR rocket grounds.

The photo dates from 1931. Immediately to the right of the rocket is its principal designer Hermann Oberth, whom we encountered earlier. Hermann Oberth wrote his book *Die Rakete zu den Planetenraümen (The Rocket into Planetary Space)* in 1923. He became technical advisor to the Fritz Lang film *Frau im Mond (Woman on the Moon)* in 1929. The design was realistic and included such aspects as waste disposal, heating and cooling. This was an early opportunity for the movie-going public to consider the possibilities of space travel. Von Braun, as we have mentioned, was subsequently recruited by the German army into making the V2 rocket as a military weapon for use during the Second World War. We have seen how, immediately following the War, the accomplishments of the German rocket team were acknowledged internationally. And in fact von Braun, with a large portion of his team under the code name Project Paperclip, (Fig 4-31) was brought to the US and recruited by the US army to lead the US in its first steps into space (see Ref 7, Ref 10).

Fig 4-31 von Braun's Project Paperclip Team members at White Sands, New Mexico in 1946 – von Braun is 7th from the right of the front row.

(Credit: Redstone.Army.Mil)

Fig 4-32 captures a powerful image of some of the key members of that team about 10 years later, with early rocket models, including the Redstone, in the background. The German leadership of the US space program is apparent. Oberth is front and center. He was at the time director of Redstone rocket production in Huntsville. Von Braun is sitting on the table at right, and von Braun's right hand man Ernst Stuhlinger, his guidance expert, is seated to the left of the image. The jet aircraft pioneer and reliability specialist Robert Lusser is to his left. Von Braun's chief, the Commander of the Redstone Arsenal, Major General Toftoy, is at back.

Name Dr Wernher von Braun

(Credit: NASA)

Summary Description
Rocket engineer and space architect

Date of Birth
March 23 1912, Wirsitz, Germany

Date of Death
June 16 1977, Alexandria, Virginia, USA

Nationality US (from 1955)/German

Achievements
Designer of V2, Jupiter, Redstone, Saturn launch vehicles. First US satellite; First US man in space. Architect behind the US Moon landings of 1969-1972.

Specific help for Your Ticket to Space
Von Braun's remarkable tale has been well documented. There is no doubt that he was the visionary behind the achievement in the sixties of the Moon Landings. He combined an extraordinary mix of engineering savvy, marketing flair, charisma, political nous and management skills, all of which was essential to bringing about the practical manifestation of his boyhood dreams of sending men into space and to the Moon. Ironically, perhaps one of the lasting legacies of the Moon Program was the photography of this planet from space. Many who have ventured into space say that the experience is transforming, and it makes such things as war seem rather pathetic. Another legacy is the widespread use of the phrase "If we can go to the Moon, then surely we can…..(complete with whatever project lies ahead)…". Now into this category comes the project of space tourism. Von Braun made everything possible, including your ticket to space.

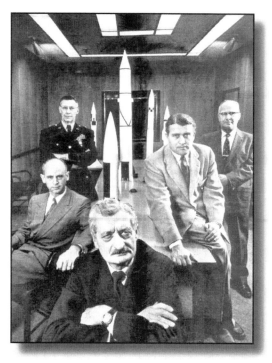

Fig 4-32 US space program management, at Huntsville in February 1956, including Hermann Oberth (front), Wernher von Braun (2nd from right), Ernst Stuhlinger (left). In the background are Arsenal Commander Major General H. Toftoy and Robert Lusser (QC Specialist).

(Credit: NASA)

The October 1957 launch of Sputnik 1 by the Soviets galvanized the US public, who demanded a response from their government. The first attempts with the civilian Vanguard rocket and satellite ended in failure. It was, however, Von Braun's army team, initially working from Fort Bliss, Texas, then moving to White Sands, New Mexico, and finally operating from Huntsville, Alabama, which got the US into space. In Fig 4-33 we see America's first successful satellite, Explorer 1, being integrated with the Jupiter-C rocket (an upgraded V2) that launched it into orbit on 31st January, 1958. We notice at this remove that in those early years the integration team worked in the open air atop the rocket. Later experience would result in clean room facilities being provided for the purpose. At last, we were getting into space. Subsequent Vanguard flights were successful, however, as we should point out in defense of the author's old Inmarsat Technical Director, and they discovered useful scientific data. In April 1959, von Braun and Goddard's widow attended the dedication of Goddard's commemorative rocket collection exhibit in New Mexico. The next step would be manned spaceflight.

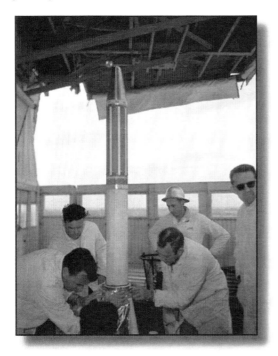

Fig 4-33 Integrating Explorer 1, America's first satellite, onto the Jupiter-C launch vehicle (Jan 1958). Notice the open "clean room" atop the upper stage of the rocket. In later years, reliability considerations would result in enclosed facilities for this critical step in preparing for satellite launches.

(Credit: www.Redstone.army.mil)

After the success of getting Explorer 1 into orbit, von Braun was tasked with building the rocket that got America's first man, Alan Shepard, into space in May, 1961. He used an upgrade of the same Jupiter vehicle, which was known as the Redstone, to launch Shepard's "Freedom 7" Mercury spacecraft (Fig 4-34). As had been the case with the Soviets, humans were not launched until animals had demonstrated the success of the life support systems. In the US, chimpanzees were used rather than the dogs that had been used by Korolev. A chimpanzee called Ham was the precursor of Alan Shepard on top of a Redstone rocket (a source of much future ribbing by Shepard's buddies, particularly the joker Wally Schirra who would say: "The animal welfare

people were concerned about the chimps, so they flew Shepard"). There had also of course been much training in withstanding the high g-forces experienced in spaceflight. Many tests of the ability of humans to withstand high-g loadings had been carried out on the ground not only by the Mercury astronauts on the training centrifuge, but by brave volunteers who rode rocket sleds, and who received very little public recognition. As we saw in the previous chapter, however, Shepard himself became a national hero with his 15-minute flight from Cape Canaveral, Florida, landing downrange along the Atlantic Missile Range. His comment was: "It's a beautiful day. Boy, what a ride!" After his flight, Shepard was treated to the same kinds of enthusiastic crowds that had greeted the earlier pioneer Lindbergh. The modern sub-orbital space tourist is in many ways repeating the mission of Alan Shepard in his Freedom 7 capsule. The weightless period was about 5 minutes, and the spacecraft did not get into orbit, but simply performed a parabolic arc, or lob. Fig 4-35 shows us the less than charming experience of the recovery from the Atlantic Ocean at the end of the flight.

Fig 4-34 Liftoff of Alan Shepard's Mercury Redstone flight on May 5th, 1961, making him America's first man in space. This flight is now the model for today's sub-orbital space tourism ventures.

(Credit: NASA)

Fig 4-35 Recovery of Alan Shepard after first sub-orbital spaceflight, 5th May, 1961.

(Credit: NASA)

Fig 4-36 and Fig 4-37 record something of the resulting acclaim after the pioneering flight of Alan Shepard. As a space tourist, even though you will in some ways be repeating Shepard's experience, you would be wise not to expect the acclaim. You probably won't meet the president, get a "Time" magazine cover, or a postage stamp, because, after all, Shepard was first and he did it about fifty years ago, before we even knew that man could survive in space.

Fig 4-36 Alan Shepard and wife Louise in Washington, DC, on their way to visit the President and Congress, following his pioneering spaceflight (1961).

(Credit: Achievement.org)

Fig 4-37 Shepard is memorialized on a Paraguayan postage stamp. There are regulations in the US preventing the use of images of living people on postage stamps.

(Credit: Author)

While you are enjoying your flight, perhaps you can try to imagine what it felt like to be the first. All the world was watching via television. Many previous unmanned launches had exploded. The US medics did not know how the human body would react to weightlessness. The Soviets were not talking to the Americans. Would there be a problem with eyesight in weightlessness? Shepard had to go into space to find out. His voice transmissions from Freedom 7 were methodical, and conveyed little emotion: "This is Freedom 7, the fuel is go. 1.2g, cabin pressure 14 psi, oxygen is go." Shepard was highly competitive and generally shared the view of his fellow Mercury astronauts that space was for military test pilots (only!) He was not therefore an advocate of space tourism (he died in 1998, just before Glenn's second flight, before the subject of space tourism was much discussed). The author met him in 1995 in San Diego, CA, and his main interest at that time seemed to be golf (he had made a golf swing on the Moon during his Apollo 14 Moonwalk, and he was by then doing charity golf competitions). He had mellowed a little by then, gave autographs, and still could muster a big boyish grin and a "Sure!" when asked for a cuddle by female admirers (which regrettably included the author's wife - what is it with astronauts, anyway?!). He little realized then how his 1961 Mercury sub-orbital flight would become the model for public space travel in the second decade of the 21st century.

So, now that we had man in space, what came next? The answer is stunning in its confidence. Less than a month after Shepard had flown his 15-minute flight into space, President Kennedy set the extraordinary challenge to the nation of getting a man safely to the Moon and back before the end of the decade of the sixties. To do this, von Braun developed the enormously powerful Saturn 5, and we see him later alongside its first stage engines in Fig 4-38. The whole vehicle was 360 feet long.

Sinatra was by now singing: "Fly me to the Moon".

Fig 4-38 von Braun at base of the Saturn 5 Moon rocket.

(Credit: US Space and Rocket Center)

Kennedy's challenge was met, over an 8-year period, involving 6 flights of Project Mercury, 10 flights of the 2-man Gemini capsule, and 11 flights of the 3-man Apollo spacecraft. Though not a consideration at the time, the critical parts of

the process leading to the Moon landing were also essential for the subsequent development of space tourism. The technology of orbital maneuvering, rendezvous and docking was perfected during Project Gemini, and used for Apollo, Skylab, the Apollo Soyuz Test Project (ASTP), and all subsequent flights to space stations. These technologies are now needed to allow orbital space tourists to reach their destination space station or space hotel. Buzz Aldrin was the astronaut whose skills most helped to master orbital rendezvous and Neil Armstrong was the first astronaut to achieve orbital docking. Fig 4-39 places us on board one of the spacecraft during a Project Gemini demonstration of orbital rendezvous. Fig 4-40 and Fig 4-41 remind us that Aldrin and Armstrong, the respective masters of rendezvous and docking, was also the chosen crew of the first lunar lander "Eagle". They arrived at the Sea of Tranquility in Apollo 11 on July 20th, 1969 (a feat that was of course celebrated throughout the world, including the Arab Gulf States where these commemorative stamps were issued).

Fig 4-39 Gemini orbital rendezvous - an essential skill for both Moon landing and space tourism. Buzz Aldrin wrote his MIT doctoral thesis on the subject, and was known as "Dr Rendezvous" to his fellow astronauts.

(Credit: NASA)

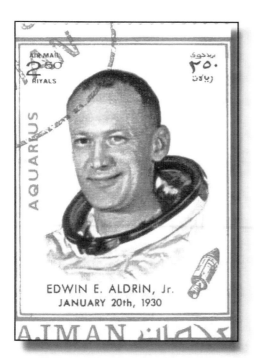

Fig 4-40 Buzz Aldrin, astronaut, master of orbital rendezvous, and Lunar Module Pilot for the first Moon landing in Apollo 11.

(Credit: Author)

Fig 4-41 Neil Armstrong performed the first orbital docking (in Gemini VIII), and commanded the Apollo 11 Mission for the first Moon landing.

(Credit: Author).

Fig 4-42 demonstrates the rendezvous activity of Apollo, with the ascent stage of the Lunar Lander "Eagle" from Apollo 11 returning to lunar orbit with Armstrong and Aldrin for rendezvous with Mike Collins in the Apollo command module "Columbia" in preparation for return to Earth. Fig 4-43 shows, after the Moon flights had been concluded, a later Skylab Space Station (1973) mission, which demonstrated how crews could survive in space for several months by visiting what was essentially an early version of a space hotel. Fig 4-44 shows the Apollo Soyuz Test Project (ASTP) (1975) mission, which demonstrated cooperation in space between the rival governments and space programs of the USSR and the USA. All of these missions relied upon the critical orbital rendezvous and docking skills which had been developed during Gemini, and which subsequently became an essential part of orbital space tourism operations.

Fig 4-42 The Apollo 11 Lunar Lander ascent stage rendezvous with the Command module in lunar orbit, July 1969. Beyond the lunar horizon is the "blue marble" of the Earth, to which the crew would return after a further 3 days of journeying.

(Credit: NASA)

Fig 4-43 The Skylab orbital station in 1973 - an early forerunner of space tourist hotels. Astronaut Pete Conrad and his crew saved this station from disaster, when they boarded and repaired it after it had been damaged during launch.

(Credit: NASA)

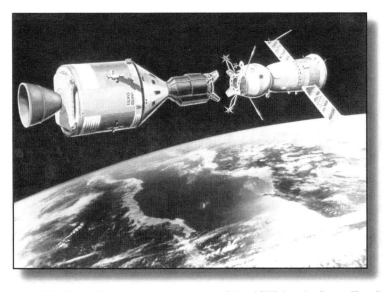

Fig 4-44 Approach and docking of the two components of the 1975 Apollo Soyuz Test Project (ASTP). This joint spaceflight ensured that internationally standardized docking equipment could be designed, which would make on-orbit rescues possible. Tom Stafford led the US part, and Alexei Leonov commanded the Soviet part of the joint operation.

(Credit: NASA)

The next major step in our progress towards making space tourism possible would be the development of a reusable launch vehicle (RLV). A truly reusable vehicle would bring enormous benefits in terms of reduced cost to orbit. Traditional expendable launch vehicles (ELV's), as were used for Vostok, Mercury, Gemini, Apollo and Soyuz were destroyed as each stage in turn was separated and left to burn up on re-entry to the atmosphere. To think about the benefits that an RLV would bring, just consider what the costs of an airline ticket would have to be, if the Boeing 747 was thrown away each time it got its passengers to their destination airport! The rocket business has needed to parallel developments in the aviation business, and become more airline-like in its operations. Fig 4-45 captures a fine view of the Space Shuttle during take-off. The Space Shuttle started flying in 1981, but was only semi-reusable, however, and its huge tank was jettisoned and destroyed each time it was launched, and its solid rockets had to be fully refurbished after recovery from the ocean whence they had landed by parachute. So it did not bring the economic benefits that had been predicted for a fully reusable vehicle, and was never used for space tourism purposes. The Space Shuttle did, however, continue to provide experience in orbital rendezvous and docking techniques, even though it just did not "deliver the goods" in terms of an airline-like operation. For the first truly reusable space vehicle, we would have to wait until 2004, and vehicles specifically designed for space tourism.

So, how do we achieve a vehicle that can get tourists into space safely and cheaply and as a matter of commercial routine? Now look at Fig 4-46. Wernher von Braun is to the right, doing one of the many things that he did so well, i.e. talking to politicians, while behind him we see one of his original rocket team, and engine expert, Konrad Dannenberg perhaps somewhat distractedly looking elsewhere. Dannenberg represents our connection with space tourism going forward. Von Braun died in 1977, Oberth in 1989, and Dannenberg not until 2009, after making many more contributions, including support for the Huntsville US Space and Rocket Center, which is an education resource where children can see the early rocket hardware.

Fig 4-45 The Space Shuttle passing through a shock wave shortly after launch. The vehicle, with its massive load carrying capability, made possible the assembly of the International Space Station, and delivery of huge payloads like the Hubble space telescope into orbit, but its operation and maintenance proved to be too complex and costly for more routine space operations.

(Credit: NASA)

Von Braun's vision, engineering and management skills had ultimately made the Moon landings possible. The author learned something of his techniques from seeing the documentation associated with his regular staff meetings. An example in the collection of the author, dated 25th October, 1965, has von Braun giving detailed instructions to his deputy about meeting a visitor to the Marshall Space Flight Center. The visitor was coming to discuss possible infrared sensing experiments during the upcoming Moon landings: "Let's put our best foot forward in these areas when he visits…" In the same management report, von Braun, with his eye for important technical detail, underlines for emphasis new data about micro-meteoroid penetration frequencies. His management skills were not confined to written work. The author talked in 2002 in Huntsville, Alabama, with one of von Braun's rocket test stand engineers, now retired and driving a tour bus. He had obvious affection for his former boss, and provided an anecdote that perhaps explains part of von Braun's success as a manager: "He was easy to talk to. He would always ask permission to come on the stand. Jeez! He ran the whole place, but still he asked us permission!" In 1965, halfway through his extraordinary and complex Mercury/Gemini/Apollo process to the Moon Landing, von Braun found time to leave Huntsville and visit San Francisco to honor the recipients of a national student design competition, organized by the American Institute of Aeronautics and Astronautics. One of the awardees, much impressed by von Braun's presence at the proceedings, was a 22 year old Burt Rutan.

Fig 4-46 Konrad Dannenberg (in background) as von Braun (right) talks to Congressmen from the House Committee on Science and Astronautics at Marshall Space Flight Center, Huntsville, March 9th 1962 (at left is engineer Dieter Grau, another Project Paperclip veteran. The Committee members are J.G.Fulton of Pennsylvania and J Waggoner of Louisiana).

(Credit: NASA)

We have already met Burt Rutan (in Fig 3-38), the designer of the SpaceShipOne vehicle which won the Ansari X-Prize for the first civilian space flight in 2004 (of which we will learn more in later chapters). Rutan was inspired by von Braun's vision, achievements and engineering skill. He had worked at Edwards Air Force Base with the X-Men (as we shall see in the next chapter), and developed skills in prototyping and making light aircraft out of composite materials. When he saw what von Braun's vision had achieved, including landing six crews on the Moon, Rutan was still a young man. He has recounted that he imagined that everybody would soon be able to experience space for themselves. But he was subsequently disappointed by the continuation, for 40 years, of the US policy of flying government astronauts only. And this motivated him to start thinking of building his own space program, using his aircraft design and composite materials expertise, starting with sub-orbital lobs.

We shall see later how he was inspired to do this. He had a small crew of dedicated young engineers, to whom he delegated lots of responsibility, and he also sought help from the experts, including our old friend. We see in Fig 4-47 a special meeting with Konrad Dannenberg talking to Burt Rutan in 2004; the topic, we may be reasonably certain, was rocket motors, the X-Prize and space tourism. After Rutan's SpaceShipOne had gone into space two times within a week in September and October 2004, he won the $10M Ansari X-Prize. Dannenberg, representing von Braun's legacy, was there to witness the event. This was the world's first truly reusable space launch vehicle. That 1965 meeting of the 22 year-old Burt Rutan and von Braun had led ultimately to a whole new vision of human spaceflight-space tourism.

Fig 4-47 Konrad Dannenberg, V2, Redstone and Saturn rocket engineer from Von Braun's team, discussing the events of the X-Prize with Burt Rutan in 2004.

(Credit: NASA)

Several other awards followed, for Rutan. He has said that his favorite was when, in a moving ceremony in Washington DC in December 2005, he received the National Space Society's von Braun Award, presented by Dannenberg. The award consisted of models of von Braun's 1952 reusable rocket concept and his 1968 Saturn 5 Apollo Moon Rocket (Fig 4-48), arranged on a plinth inlaid with black Italian granite. Dannenberg voiced his support for space tourism at various conferences and media interviews, and the author heard him in 2007 stressing the importance of quality control, especially in soldering and welding, in making this possible.

Name Konrad Dannenberg

(Credit: Author)

Summary Description
German rocket scientist, part of von Braun's Team.

Date of Birth
August 5 1912, Weissenfels, Germany

Date of Death
February 16 2009, Huntsville, Alabama, USA

Nationality US/German

Achievements
Engineer on V2, Redstone, Jupiter, Saturn 5. US Space and Rocket Center supporter.

Specific help for Your Ticket to Space
Konrad Dannenberg had seen all of the developments of rocketry from the World War II weapons, through the space race and Moon landings of the sixties, to the arrival of space tourism. He was an enthusiastic advocate of educating the young to the benefits of space technology, and gave freely of his time at the Huntsville US Space and Rocket Center for that purpose. A space tourism booster, he supported Burt Rutan's work on SpaceShipOne at the time of the Ansari X-Prize in 2004.

Fig 4-48 Burt Rutan's von Braun Award for designing SpaceShipOne, which made the first civilian space flight in 2004, thus opening up the space tourism industry.

(Credit: National Space Society)

Fig 4-49 Konrad Dannenberg presents National Space Society's von Braun Award to Burt Rutan, while Buzz Aldrin looks on, Washington DC, 2005.

(Credit: Author)

Also present in the presentation event shown in the Fig 4-49 image, right in the foreground, is another key player in bringing space tourism into existence, Buzz Aldrin, whom we noted in 1969 was the Lunar Lander Pilot of the first Moon landing in Apollo 11. He, it was, who speaking from the Moon defined the view of the Sea of Tranquility as: "Magnificent Desolation!" As we have seen, Aldrin, with a Doctorate of

Science in Astronautics from MIT, had developed the practical skills to carry out rendezvous and docking in orbit during Project Gemini and Apollo. These skills are now essential for getting space tourists to their space hotel. We shall hear more of his contribution later. Konrad Dannenberg died in February 2009, aged 96, and among his many other achievements, he had made major continuing efforts at educating children in the rewards and benefits of space flight and space tourism. Fig 4-50 and Fig 4-51 testify as to how the early Mercury astronauts Alan Shepard and Wally Schirra appreciated Dannenberg's work. When Shepard was ready to go on the first US manned spaceflight, and he was frustrated waiting on top of the rocket for all of the safety checks to be completed, he famously called over the intercom: "Let's light this candle!" Dannenberg had been responsible for ensuring that the Redstone engines of "this candle" gave Shepard a safe ride.

Fig 4-50 Alan Shepard acknowledges the contributions of Konrad Dannenberg in Huntsville, November 1986. Dannenberg helped Shepard walk on the Moon.

(Credit: Huntsville Times)

Both Shepard and Schirra also rode the Saturn vehicle, whose engines Dannenberg had designed. Schirra, as we have previously noted, was a famous joker, and he used to tell jokes about the German leadership of the US space program (Ref 7), and in fact about the reliance on German engineering talent on both sides of the "space race" between the US and the USSR. Von Braun, himself, is reported to have enjoyed one of them in particular, asking Schirra to repeat it more than once. Wally Schirra's story was that a Soviet and

a US spaceship landed side by side on the Moon. The two commanders descended their ladders and loped across the lunar surface in one-sixth gravity to greet each other. In his mock-German accent, Schirra then recounted their conversation: "Hello Fritz. Hello Hanz. Now we can both speak in German!" Wally Schirra died in 2007, but never really embraced the idea of space tourism. As one of the original Mercury Seven astronauts, he still regarded space as the proper preserve of military test pilots.

Fig 4-51 Project Mercury Astronaut Wally Schirra and Konrad Dannenberg at a 2005 space conference. Schirra underlined the essential contributions of the German rocket engineers.

(Credit: Wallyschirra.com)

We have now watched the chain of developments in rocketry, and subsequently commercial space, right up to the point where space tourism becomes possible. But it was the Russians who were the first to take the final step, following the vision of Tsiolkovsky. Let's return to Russia, where we started this chapter. The very first space tourists (see Table 2-1) flew using Russian vehicles, and they flew on the latest upgrade of the Vostok rocket (called Soyuz) that Korolev had designed, and which launched the first satellite and the first human into space. The former Soviet Union was in effect teaching the West how to bring capitalism into the space environment. You may recall that they were learning how to do this during the Soviet era at Inmarsat in the early '80s. They began their space tourism activities at the end of the Soviet era, and have continued ever since. They had no problem, even as a formerly communist society, taking lots of hard currency from western millionaires, in return for a flight in an otherwise empty third seat of a Soyuz capsule. Korolev had died in 1966, just before the test flights of his new Soyuz spacecraft. That same spacecraft flies today, 44 years later, and not only services the International Space Station with its occupants and supplies, but also has been the space tourism vehicle used by absolutely all space tourists to date.

Fig 4-52 places us with the space tourist Anousheh Ansari a few days before boarding her Soyuz rocket for an orbital flight following almost literally in the footsteps of Gagarin. He had gone up the same stairway on the same gantry to his Vostok capsule in April 12th 1961. Back then, the launch site was a closely guarded secret, and was within the territory of the former Soviet Union. Ansari's launch took place Sept 18th 2006, from the same departure point that Gagarin had used for the first manned flight into space, but now the Baikonur launch complex is geographically within the territory of the independent sovereign state Kazhakstan. The modern orbital space tourist uses many of the same facilities and training equipment that were used by the early Soviet Vostok cosmonauts.

The "Wright Stuff" was clearly in evidence wherever the Rocket Men (and Women) worked to turn their dreams into reality. To single any one out for the coveted award might seem to be invidious, so we could be excused for making a collective presentation: To All of the Rocket Men. However, we must not shirk our responsibility, and can only spare two awards, so we give the accolade for the Rocket Men category of the "Wright Stuff" awards to Sergei Korolev and Wernher von Braun. They had the dreams, but also took the risks to make the future happen. Space Tourism benefited enormously from their endeavors. We might also give a special tip of the hat to Arthur C Clarke and Konrad Dannenberg for adding their own distinctive contributions.

While contemplating your upcoming spaceflight, consider the courage and determination of those rocket men and astronauts who led the way towards making your ticket a possibility. We now need to see how the extraordinary exploits of the aviators and the rocket men came together.

Fig 4-52 Iranian/American citizen Anousheh Ansari points to the Russian Soyuz rocket for her September 2006 orbital space tourism flight. She was the second woman, after the British Helen Sharman, to undergo this experience of being a commercial private passenger in a Soyuz.

(Credit: Space Adventures)

Sergei Korolev

Wernher von Braun

"One of the phrases that kept running through the conversation was "pushing the outside of the envelope". The "envelope" was a flight-test term referring to the limits of a particular aircraft's performance. "Pushing the outside", probing the outer limits, of the envelope seemed to be the great challenge and satisfaction of flight test."

Tom Wolfe, "The Right Stuff", 1979 (Ref 1)

We have seen how aviators provided the base that was essential to space tourism. We have seen how rocket engineers and astronauts were essential to getting space tourism off the ground. These disciplines in a sense come together within the profession of the rocket plane test pilots, whom we will call the X-Men, since most of their aircraft had an X-designation, to represent their experimental nature. This chapter will look at a part of the space story that might seem to be a diversion from the main developments, but we shall see that it is anything but that. It is mostly about the test pilots of the so-called X-Planes, the experimental rocket planes that fly from Edwards Air Force Base in the high desert north east of Los Angeles in California. The Mojave Desert is their home territory. When Tom Wolfe wrote his insightful book: "The Right Stuff" (see Ref 1), while it told the tale of the Mercury astronauts, its real hero was Chuck Yeager who first broke the sound barrier in his Bell X-1 rocket plane in 1947.

But first we need to recognize the work and courage of earlier rocket plane pioneers who were not operating from the US. The first attempts to use rockets to power aircraft were conducted by the Austrian Max Valier (Fig 5-1) and Fritz von Opel (Fig 5-2) in Germany in 1929. During the late 1920's, von Opel and Max Valier had been driving rocket-powered racecars. They then converted a sailplane by attaching 16 solid rockets, called it the Opel RAK-3, and it was flown for 75 seconds, covering a mile, but was destroyed on landing (Fig 5-3). Valier had read Oberth's book in 1923, and he wrote his own book in 1924 "*Der Vorstoss in den Weltenraum (Advance into Space)*", advocating an evolutionary program of rocketry, starting with cars and sledges, moving on to gliders and airplanes, and leading ultimately to a rocketship. Valier was killed in 1930 when one of his rocket engines exploded. Konrad Dannenberg's interest in rocketry had begun when he heard Max Valier give a lecture.

Fig 5-1 Rocket plane pioneer Max Valier.
(Credit: American Institute of Aeronautics and Astronautics)

Fig 5-2 Fritz von Opel (aged 29) in his rocket car powered by 24 solid fuel rockets on May 23, 1928.

(Credit: cars-spot.com)

Fig 5-3 The first Rocket plane – the Opel RAK-3 (1929).

(Credit: Absoluteastronomy.com)

Then, in 1944, Allied pilots during the Second World War were mightily surprised to meet up with the Messerschmitt ME 163 Komet (Fig 5-4), which was the world's first operational rocket plane fighter, and could fly at almost 600 mph. It had of course no propeller, and its wings were swept-back, and it looked most unlike anything the Allied pilots would have seen. Fortunately for the 350-mph Spitfire pilots, the ME 163 could only carry enough fuel for eight minutes of powered flying. It was nevertheless able to shoot down RAF Lancaster bombers and USAF B17's.

But, as we know, Germany lost WWII, and Britain was severely weakened, so the action moves to the US, and in particular to the Mojave Desert. In Tom Wolfe's tale about the early

US rocket and space flights, "the right stuff" referred to the essential characteristics of the test pilot. And to Wolfe, Chuck Yeager was the embodiment of the term. In Wolfe's telling, there was a significant rivalry between the men flying the rocket planes in the skies over Mojave, and the new breed of astronauts who emerged at the beginning of the 'sixties. Yeager's expressed view was that it needed more skill to fly rocket planes than to be an astronaut in a capsule, or "spam in a can", to use his own colorful term. The Mercury astronauts, however, were to take all the newspaper headlines. The exploits at Edwards were sidelined by the developments of the space program which started with Mercury, and went on via Gemini to the Apollo Moon landings at the end of the decade of the 'sixties. We shall see that the story of rivalries between the X-Men and the astronauts perhaps overstates the reality, and that it was rather the case that Edwards had a significant role in the subsequent emergence of space tourism. Your own upcoming space tourism flight owes much to the work and courage of the X-Men.

Fig 5-4 The Messerschmitt Me 163 Komet operational rocket-powered fighter (1944).

(Credit: Museum of Flight, East Lothian, Scotland)

Let's start with Chuck Yeager himself (Fig 5-5). It is perhaps difficult for a modern young reader to appreciate, but when jet planes first became available, towards the end of the Second World War, there seemed to be a very real barrier to their speed as the speed of sound was approached. Some test aircraft literally fell apart when they met the "Sound Barrier", as it became known. It was all a matter of getting the aerodynamics right, and finding brave test pilots to try out the new designs. Of course, it goes almost without saying, that you cannot have space tourism, or much of a space program at all, unless you can figure out how to fly supersonically. Yeager was the man to do it. He was only 24 when he broke the sound barrier. The rocket planes were carried aloft, slung underneath a converted USAF bomber. At altitude, the rocket plane was dropped from the mother plane, the rocket engine was ignited, and the X-plane headed upwards. The rocket motor only had limited duration, so the X-planes landed as gliders on the enormous dry lakebeds at Edwards. Neither the Bell X-1, Yeager's ride, nor the subsequent X-planes, looked much like gliders – and they did not fly much like gliders – and the X-Men needed great skill to land safely after each mission. Skills that the pilot/astronaut of your own space tourism craft must have also.

Fig 5-5 Chuck Yeager beside the cockpit of his Bell X-1 rocket plane in 1947.

(Credit: USAF)

The early test flights were secret, and when Yeager broke the sound barrier, just barely exceeding Mach 1, on October 14th, 1947, there were no journalists to record the feat; the news was announced much later. But Yeager was already an important American figure. He was a war hero from the Second World War, even having taken part in clandestine operations, and he brought test flying to the attention of the American public (see Ref 2, Ref 8). The Bell X-1 "Glamorous Glennis", named after Yeager's wife, was presented to the Smithsonian in 1950. The aircraft had made 83 flights during its research activities. Yeager was able to eventually "push the envelope" in the X-1 to Mach 1.45, which he achieved in March 1948.

Name Chuck Yeager

(Credit: National Air and Space Museum)

Summary Description
Military test pilot

Date of Birth
February 13 1923, Myra, West Virginia, USA

Date of Death
n/a

Nationality US

Achievements
Flew Bell X-1, the first aircraft to fly supersonic, in 1947.

Specific help for Your Ticket to Space
Chuck Yeager flew the first US generation of craft that combined the attributes of aircraft and rockets. In so doing, he not only was the first man to exceed the speed of sound, but he paved the way for the first space tourism craft, such as SpaceShipOne and SpaceShipTwo. In Tom Wolfe's book "The Right Stuff" about the Mercury astronauts, Yeager was described as the embodiment of the test pilot, although he never himself became an astronaut. It is extremely doubtful whether Yeager ever actively encouraged the idea of space tourism; he belonged to the old school that stressed the need for combat-trained veteran pilots to tackle the unknowns of high altitude and high speed flight. General Yeager and his successors have, however, gradually removed those unknowns – and some of his colleagues paid the ultimate price for that knowledge. So now it has ironically become possible for the general public to go higher and faster than the general himself.

In Fig 5-6, we see Yeager meeting President Dwight D. Eisenhower at a White House award ceremony on November 23rd, 1954. He is sharing the honors with the pioneering woman jet pilot Jacqueline Cochran, as they collect their respective Harmon Trophy awards. Cochran had created the Women's Airforce Service Pilots (WASP's) organization used to ferry aircraft during WWII from the manufacturers to where the fighting was taking place. The WASP's eventually received congressional recognition, and the 80- and 90-year old survivors received the Congressional Gold Medal in March 2010, just as this book was going to the publishers. Cochran had become the first woman to break the sound barrier in 1953.

Fig 5-6 Chuck Yeager and Jacqueline Cochran and President Dwight D. Eisenhower at the 1954 award ceremony for the Harmon Trophy.

(Credit: Bettmann/Corbis)

In 1955, when the president was trying to decide what to do about space in the post-war world, he talked with both Yeager and Lindbergh. The idea of artificial satellites had been discussed, and during WW2 Eisenhower had seen what his adversaries could do in terms of developing rocketry. Could satellites in some way be used as offensive weapons? We have seen that the US had even transferred a whole team of the former enemy Germany's rocket engineers, under the code name Project Paperclip, to White Sands in New Mexico to pass on their skills to the US. Charles Lindbergh, as we have also seen, had been following and supporting the results of Goddard's work into liquid rocket technology in a neighboring part of the New Mexico desert. Yeager, we have just learned, had been going higher and faster in the rocket planes at Edwards,

with space as the ultimate target. Which of these approaches promised the best way forward for the US in the attempt to reach space? What we know, in retrospect, is that the main thrust of developments during the subsequent "space race" with the Soviets followed the von Braun/ Goddard/ Korolev vertical multi-stage rocket approach, which eventually led to the Moon landings. However, we shall also see that the rocket plane approach, pursued by Yeager and his fellow X-Men at Edwards, would come into its own much later, and even provide the mechanism for the first sub-orbital space tourism flights.

Even though the Mercury astronauts had taken center stage, Yeager's time in the spotlight was not over, because ironically he was later hired as an advisor in the making of the movie of Tom Wolfe's book in 1982. He even played a cameo part in it as a bartender at the reconstructed Pancho Barnes' Happy Bottom Riding Club used as a set location. The original bar, in the Mojave Desert just outside Edwards Air Force Base, where the early X-Men used to congregate and discuss their latest exploits, had been destroyed by fire. The bar was celebrated for having a wall of photos of famous test pilots who had lost their lives pushing the envelope of high altitude, high speed flight, in the skies above Edwards.

Yeager, in fact, flew with many of the astronauts at various times. He worked with Gus Grissom when that Mercury astronaut came out to Edwards in October 1962 to test fly some landing techniques for the upcoming Gemini missions using a special vehicle called the Parasev. He flew with Neil Armstrong, and Dave Scott, both of whom were among the 12 men who walked on the Moon with Project Apollo.

Fig 5-7 captures him with Mercury astronaut Gordo Cooper in later years, and Fig 5-8 has him with Apollo Moonwalker Dave Scott after Yeager had just celebrated his supersonic feat with a repeat performance at age 85. Cooper was noted for his precision flying – "Right on the old bazoo" being one of his phrases, and through family connections he had even flown with Amelia Earhart as a child. The author recalls hearing Cooper, at a Mercury celebration event in 2002, describe in his spare, slow Oklahoma drawl, what it was like to fly the Atlas: "You remember that great feeling when you get into a fast car and step on the gas and it accelerates?.......Like that....... Then it goes on...... and on...... and on...... Then pretty soon you get to thinking maybe you've made a serious error of judgment!" One would suppose that Yeager would totally understand such things. As recently as 2009, however, at the 40 year anniversary of the Moon landings, he was still "Harrumphing" to the press and saying that money was wasted on the Apollo program!

in newsmagazines and both had successful autobiographical books to their name (see Ref 2, 3). They both used to carouse at the bar at Pancho's (and Crossfield's character also appears in the movie of "The Right Stuff"). Crossfield was the first man to fly at twice the speed of sound, which he did on the 50th anniversary of the Wright Brothers' first flight in 1953 in his Douglas Skyrocket. Crossfield then went on to design and fly the North American X-15 (Fig 5-10), which became the X-Plane which flew faster and higher than all its predecessors, and ultimately even reached the fringes of space (see Ref 4, Ref 5, Ref 6, Ref 7). This became the model that Burt Rutan would follow when he designed his own space program. Rutan designed his own mother plane, White Knight, to carry his own spacecraft, SpaceShipOne, to drop altitude. SpaceShipOne was then released, its motor ignited, and the craft headed for space. It then returned to Earth as a glider.

Fig 5-7 Mercury astronaut Gordon Cooper (left) with Chuck Yeager, around 2001.

(Credit: Chuckyeager.com)

Fig 5-8 Yeager and Dave Scott (Apollo Moonwalker) in 2007, after Yeager's celebratory 60-year supersonic flight.

(Credit: Chuckyeager.com)

Fig 5-9 Rocket plane test pilot Scott Crossfield.

(Credit: EAA.org)

Of course, Chuck Yeager was only one of the X-Men at Edwards. Almost as famous at the time was Scott Crossfield (Fig 5-9). They both appeared

Fig 5-10 Scott Crossfield with the X-15 rocket plane he designed and piloted, at Edwards Air Force Base, Mojave, California, 1959. The X-15 program's operational approach was a guide to Burt Rutan as he developed the SpaceShipOne program.

(Credit: Allan Grant)

Fig 5-11 Scott Crossfield (left) and Neil Armstrong (right) at Edwards Air Force Base with X-15s on Feb 9th, 1961. Also at center in the photo is another X-15 pilot, USAF Major Robert White.

(Credit: Bettmann/Corbis)

In 1951, Crossfield also had a major part to play in the design of what became popularly known as spacesuits, and you can see him wearing an early variant in the Fig 5-10 image. His testing with the pressure suit, and even his decision to make the outer layers silver, almost defined the public's view of astronauts, when the Project Mercury team started flying their sub-orbital missions in 1961, a full decade later. In Fig 5-11 we capture Scott Crossfield at Edwards Air Force Base, standing at left against the nose of an X-15, handing over the symbolic keys. At the right of the photo, leaning on another X-15, is Neil Armstrong, subsequently the first man on the Moon.

Fig 5-12 lets us see Armstrong installed in the cockpit of his X-15. He was the only one of the Apollo astronauts to have had the benefit of flying the rocket plane before transferring to capsules on rockets launched from Cape Kennedy. Armstrong flew the X-15 rocket plane 7 times, reaching an altitude of over 200,000 feet, between November 1960 and July 1962, before joining NASA's astronaut corps to fly Gemini and Apollo spacecraft.

Fig 5-12 Neil Armstrong in the X-15 cockpit at Edwards Air Force Base in 1961.

(Credit: NASA)

In his Gemini VIII flight with David Scott, in March 1966, Armstrong was glad of the experience that he had gained with X-15 high altitude thrusters, when he encountered a problem. One of the Gemini thrusters stuck in the open condition, causing the capsule to rotate rapidly out of control. He aborted the mission, and brought the capsule and crew down safely to an emergency landing.

Name Albert Scott Crossfield

(Credit: Author)

Summary Description
Rocket plane test pilot

Date of Birth
October 2 1921, Berkeley, California, USA

Date of Death
April 19 2006, in air crash, Georgia, USA

Nationality US

Achievements
Designer of X-15. First to reach twice the speed of sound

Specific help for Your Ticket to Space
Scotty Crossfield was a much-loved member of the test pilots' elite. His major contribution was to come up with the concept, and manage the engineering and early test flights, of the X-15 rocket plane, which eventually exceeded the altitude of 100km, defined internationally as the boundary of space. Burt Rutan used the performance of the government - funded X-15 as his guide and challenge as he developed the commercial SpaceShipOne for the Ansari X-Prize in 2004, and he was able to exceed the altitude achieved by the X-15 almost forty years earlier. Crossfield can surely be said to have had both the "Right Stuff" and the "Wright Stuff" – since he trained the pilots who flew the reproduction Wright Flyer for the Centennial of flight.

Fig 5-13 Scott Crossfield works on Wright Flyer replica for the December 2003 Centenary of first flight.

(Credit: EAA.org)

We can see that Scott Crossfield is another of those folk with the "Wright Stuff" who can link us both backward in time and forwards to the space tourism era. Fig 5-13 effectively demonstrates this by letting us see him in 2003 working on a replica of the Wright Flyer.

Crossfield was responsible for oversight and pilot training in building and test flying the replica, in order for it to be flown in December 2003 for the centenary of the first flight at Kitty Hawk. Test flying the replica was quite hazardous, because modern flight controls had evolved in different directions to the way envisaged by the Wrights (as we saw in Chapter 3). Also, wind speed was critical to success, as the Wrights had known when they selected the Outer Banks of North Carolina for their early experiments. Crossfield learned, by hard experience, just how it had felt for Orville Wright at the onset of the aviation era. The commemoration flight subsequently took place successfully, so Scott Crossfield had thereby both honored the Wright Brothers' 50-year

celebration with his Mach-2 flight, and their 100-year celebration with the re-flight of the Wright Flyer. The author watched Crossfield give a talk on November 20th 2003, while outside the lecture room at the Smithsonian were suspended three of his famous "rides": the Douglas Skyrocket, the North American X-15, and the 1903 Wright Flyer. It was a double celebration. On that day it was 50 years since Crossfield became "the fastest man alive" by flying at Mach 2 in the Skyrocket. And it was the first day that his team had gotten the 1903 Flyer into the air, 100 years on from Kitty Hawk. The event program summed it up beautifully:

Douglas Skyrocket – Mach 2.005
North American X-15 – Mach 6.72
1903 Wright Flyer – Mach 0.04

As we have noted earlier, one of the flight test engineers at Edwards Air Force Base, between 1965 and 1972, was a young man called Burt Rutan (Fig 5-14). He was not an X-plane pilot; his job involved flying in the back seat of F-4

Phantoms to understand their stall characteristics, and then write up the flight procedures to avoid the problem. However he did have dreams of his own about going into space. Big dreams.

Fig 5-14 Burt Rutan, master aerospace engineer and space tourism visionary.

(Credit: Scaled.com)

Rutan subsequently left Edwards, and set up shop in 1974 as a designer of lightweight kit planes and based himself at Mojave airport, just along the desert road from Edwards Air Force Base. In that same year Charles Lindbergh died. We shall see later how Rutan designed the first civil craft to reach the boundaries of space, which he called SpaceShipOne. And the chosen launch method, and even the craft's shape, has a lot in common with Chuck Yeager's X-1. In determining whether the craft had in fact reached the 100km (i.e. 62 miles) target altitude for the Ansari X-Prize attempt, Rutan even called upon the radars of his old employer at Edwards for verification. He was, of course, glad to win the $10M Ansari X-Prize for his feat, but possibly even more satisfied by surpassing the X-15 altitude record in doing so. Without the resources of the US government and the Air Force, he had created his own space program with spacecraft that went higher than those of his former US government employers. Burt Rutan, the young

test engineer for the X-Men, had become the man to open up the world of space to ordinary citizens. Scott Crossfield was there in witness, and to congratulate Rutan's test pilot Mike Melvill on his flight (Fig 5-15). After the success of the Ansari X-Prize flights, Rutan took SpaceShipOne and its carrier vehicle White Knight (which he had also designed) across country, and Fig 5-16 offers an appropriate memento from that cross-country trip.

Fig 5-15 Mike Melvill and Scott Crossfield discussing test flying rocket planes, following the successful X-Prize flight, Mojave Spaceport, California, Sept 2004.

(Credit: Don Logan)

Fig 5-16 Meeting of X-Men. Mike Melvill, test pilot of Scaled Composites and pilot of SpaceShipOne, and Scott Crossfield at Oshkosh, Wisconsin, July 2005.

(Credit: EAA.org)

At the stop *en route* at the Oshkosh, Wisconsin, Air Show of the Experimental Aircraft Association, we see in Fig 5-16 Scott Crossfield, sharing the stage, and in animated conversation yet again with Mike Melvill, the test pilot of SpaceShipOne who flew two of its three flights into space. They had much to discuss. They were both experimental rocket plane test pilots. For both of them, they looked down from their apogee to the brown, purple, white and red hues of the Mojave Desert. Crossfield died a year later when the aircraft he was piloting crashed during a thunderstorm. Scott Crossfield last flew the X-15 in 1960; Melvill flew SpaceShipOne into space in June 2004 just a year after the Wright Centenary, thereby opening up the era of space tourism 44 years later. Melvill received the first civilian pilot astronaut wings from the FAA for his feat. And Crossfield had even made the Wright Flyer take to the air again in 2003. The author heard him describe his experiences in Ohio, the birthplace of the Wright Brothers, at the commemoration of a century of flight. His fellow speaker was John Glenn, who was also from Ohio, and who used the opportunity to express his concerns about the decline in math and science education in the US. This was almost full circle back to when Kennedy called for the improvements needed in these skills at the dawn of the space age. It had been quite a century. The X-Men had played their part, and the benefits would continue into the new phase of the space tourism era (provided that the US education system can generate the future scientists and engineers that will be needed).

We award Scott Crossfield and Mike Melvill our "Wright Stuff" awards for the X-Men. Yes, Melvill deserves to be included as one of the X-Men, and he certainly was crucial to the fact that you now have a ticket to space. We shall hear more of Melvill in Chapter 8.

Scott Crossfield

Mike Melvill

"We shall send to the Moon, 240,000 miles away from the control station in Houston, a giant rocket more than 300 feet tall, the length of this football field, made of new metal alloys, some of which have not yet been invented. Capable of standing heat and stresses several times more than have ever been experienced. Fitted together with a precision better than the finest watch, carrying all the equipment needed for propulsion, guidance, control, communications, food and survival, on an untried mission, to an unknown celestial body, and then return it safely to earth, re-entering the atmosphere at speeds of over 25,000 miles per hour, causing heat about half that of the sun - almost as hot as it is here today(!) - and do all this, and do it right, and do it first before this decade is out."

President John F. Kennedy, Rice University, TX, September 12th, 1962

We have already seen how aviation and spaceflight were important to US presidents (we saw Johnson with Lindbergh and Eisenhower with Yeager). But in fact such involvement goes back to the very outset of aviation. Theodore Roosevelt (1901-1909) and William Taft (1909 – 1913) feted the Wrights; Calvin Coolidge (1923-1929) celebrated Lindbergh's achievement (Fig 6-1), awarding him the Medal of Honor (see Ref 1). But, did any of the procession of presidents have any real impact on making space tourism happen? Not many. But we shall be able to make at least one "Wright Stuff" award! One thing we know for certain; there can be no space tourism without the support of government. And the president has a key role in ensuring that the support of the public and Congress is there, and ultimately that the sustaining regulatory environment is created.

Fig 6-1 President Calvin Coolidge greets Charles Lindbergh at the White House, June 11th, 1927.
(Credit: Lindbergh Picture Collection, Yale University)

The behavior of presidents in celebrating aviators and astronauts says a lot about politicians in general, and about the American ethos. Aviators had certainly captivated the American public; and of course politicians had better be aware of just what interests their voters. From early on, it became evident that aviation would bring about a new phase in military endeavors, and so budgets began to be appropriated for this purpose. During Woodrow Wilson's presidency (1913-1921), aircraft (which were mostly European designs) took part in the First World War, in 1914-1918, which was only a dozen years after the first flight of the Wright Brothers. At the end of the war, the Congress made surplus aircraft available for training and airmail. Wilson, furthermore, created the National Advisory Committee for Aeronautics (NACA) in 1915, which was the pre-cursor organization to NASA (see Ref 2), and which developed technologies such as standard aerofoils and streamlined engine cowlings. Herbert Hoover (1929-1932) made his most significant contribution before becoming president, when as Secretary of Commerce he introduced the Kelly Air Mail Act, whose importance was explained in Chapter 3.

A succession of US presidents has supported aviation developments, which gave FDR (1933-1945) and America, a decided lead in aviation up to the Second World War (1939-1945), and enabled the subsequent flourishing of the airline industry. On the aviation side of events, after World War 2, as we saw in Chapter 3, again there was a surplus of aircraft, and many new airlines were formed using surplus Douglas DC3 Dakota transports for their fleet. Prices began to be accessible to the general public, and package tour holidays were introduced, with hotel and airfare all-inclusive. The jet engine was simultaneously invented in Germany and in the UK during the war, but the US soon caught up and, as we have just seen, via its X- plane activities at Edwards Air Force Base rapidly took the lead again in pushing the envelope. There was not much political risk to the early presidents in celebrating the exploits of the first aviators. They tended to get involved after the flight had been successful, rather than before it had taken off. Few American voters would have objected to spending to bring about the new aviation industry, to provide a strong air force, and ensure that US aircraft remained in the forefront of aviation technology. In 1948, Orville Wright died. He had, in his lifetime, witnessed everything from the first flight to transcontinental flight, to passenger air travel and on to supersonic flight. And he had seen the beginnings of rocketry.

It would not be such a simple issue for presidents once the space frontier was opened. A succession of presidents and politicians has had to view a balancing of the political and financial capital against the risks of the space program. We shall see that an unlikely combination of presidents has done its part to bring America from the dawn of space travel to the onset of space tourism. Although all of them were willing to use photo opportunities with astronauts to help solidify their own personal popularity; we shall find that only two of the presidents really deserve to have the "Wright Stuff" label applied. Mainly because of their willingness to take significant personal political risks in advancing the cause.

Fig 6-2 records then former President Truman (1945-1953) looking at models of the arsenal of rockets that were under development by 1961. Truman had made the decision to use the Atom Bomb over Japan to end the Second World War in 1945, and after the war he was concerned to ensure that the US had the missiles that he thought were needed to deliver atomic warheads over intercontinental ranges.

Amongst the rockets are Jupiters, Thors,

Atlases, and some future concepts. At this time, these vehicles were known as Inter-Continental Ballistic Missiles (ICBM's), and had been developed to carry nuclear warheads. Some adaptations later turned them into satellite launch vehicles, and some were man-rated for human spaceflight.The warheads were removed and astronauts in their capsules would take their place. As John Glenn once jokingly observed: "You want me to do WHAT??"

Fig 6-2 Former President Truman views missile projects with NASA Administrator James Webb in 1961.

(Credit: NASA)

We have talked about President Dwight Eisenhower's early concerns, and how he co-opted advice from Chuck Yeager and Lindbergh. He was president from 1953 to 1961. As it turns out, the Soviets would launch their first satellite, Sputnik 1, in October 1957, but Eisenhower could not have known that at the time. In fact, even when the launch of the first satellite happened, he was not very concerned about the event. However, the American press and public's reaction was very different, and there was strong pressure for the US to show its own technological prowess in the space arena. Eisenhower, changing his attitude, put in place the necessary legislative steps, and NASA was created on July 29th 1958. It is amazing to recall that only 11 years later, they would put a man on the Moon.

NASA was created from some pre-existing facilities and teams, including the NACA institution that Woodrow Wilson had established in 1915. The President had been working on an approach to develop a US satellite, and he had insisted that the effort would be scientific, rather than military, and so he had linked it to the International Geophysical Year. A special civilian launch vehicle, called Vanguard, with no heritage as a missile, was being developed for the purpose. In doing this, he chose to put to the side the existing rocketry that the US Army had in hand, including the Jupiter that von Braun had developed from his successful V-2 vehicle. One can only speculate in hindsight why Eisenhower took this position, as we look at Fig 6-3, with Eisenhower and von Braun later at the opening of NASA's Marshall Space Flight Center in Huntsville, Alabama. It had been converted from the Army Ballistic Missile Agency, and von Braun was its first Director.

Fig 6-3 President Eisenhower and Wernher von Braun in 1960 at Marshall Space Flight Center in Huntsville, Alabama.

(Credit: NASA)

But at least in the view of the author, it seems possible that Ike, the former WWII Supreme Allied Commander, may not have wanted to put so much trust in the hands of his former German World War II adversary. However, the result was that the Soviets got off to a good start in the "Space Race" of the 'sixties, and they produced in fairly quick succession a number of firsts. They launched the first satellite (1957), the first living being into orbit (the dog Laika, also in 1957), the first probes to the Moon (1959), the first man in space (Gagarin, April 1961), the first man to spacewalk (Leonov, 1965). Meanwhile, the first attempts of the US to get something into orbit using the Vanguard rocket were spectacular failures (despite a lot of late hours worked by Marty Votaw and his colleagues!). Eventually, von Braun was given the call to try and launch a satellite, and he, as we have seen, succeeded with Explorer 1 on January 31st, 1958, launched aboard his Jupiter–C rocket.

The next president, who was elected in 1961, had a very different view of space travel, and its importance to America. A journalist asked John F. Kennedy, at an early press conference, whether the US could get to the Moon before the Russians. His answer was immediate: "If we can get to the Moon before the Russians, then we should." The new NASA Administrator, appointed by Kennedy, who would have to make this happen was James Webb, who held the post from 1961 through to 1968. The Mercury program had been established in 1959, and, as we noted earlier, Alan Shepard became America's first man in space in May 1961. That flight had not gone into orbit. It was a sub-orbital lob in the Mercury Capsule "Freedom 7", on board a von-Braun derived vehicle called the Redstone, and lasted only 15 minutes (a flight in its profile and duration which was very similar to the SpaceShipOne flights of 43 years later).

But this was what the US public wanted to see. Alan Shepard was paraded as a public symbol of the US technological achievement of space flight. Kennedy, as a superb politician, immediately recognized the public interest and support. Even so, his next step was extraordinary for any politician to take, and it is the reason he gets one of the two "Wright Stuff" accolades for US presidents. He committed the country to the incredibly expensive and risky challenge of getting a man on the Moon, and returned safely to Earth, before the end of the decade of the 'sixties. Kennedy's rallying call evoked all of the best qualities of being American; accepting a challenge, taking risks, pushing boundaries, courage and creativity. Because of his extraordinary decision, all of the technologies and skills were developed that would later prove to have been essential for the emergence of a space tourism business, and arguably a half a century before it would otherwise have happened. We know in retrospect that the feat was accomplished within 8 years, but what an audacious gamble to take when the country had at the time only 15 minutes of accumulated experience of manned space flight. The space race was underway. All subsequent astronauts would be treated as public symbols of achievement, and as such would be courted by presidents. In Fig 6-4 we observe the super-salesman von Braun with President Kennedy.

Fig 6-4 President John F. Kennedy and Werner von Braun in May 1963.

(Credit: NASA)

Fig 6-5 President Kennedy and Mercury astronauts in 1963. (Virgil Grissom, Gordo Cooper, Al Shepard, Wally Schirra, Scott Carpenter).

(Credit: NASA)

As a former young naval officer, Kennedy enjoyed being around the Mercury astronauts, who were also military officers, and hearing their stories first hand. He probably would have liked to go into space himself. We know for certain that this was the case with von Braun. Fig 6-5 shows President Kennedy amongst five of "his" Mercury astronauts: Gus Grissom, Gordo Cooper, Al Shepard, Wally Schirra and Scott Carpenter. Behind him, and to his right, was his Vice President Lyndon Johnson. Cooper flew the last Mercury mission, 22 orbits in the "Faith 7" capsule, in May 1963.

Lyndon Johnson had been JFK's main advisor on space matters when he decided to set course for the Moon. He would take over as president after Kennedy's assassination on November 22nd 1963. John F Kennedy never saw the culmination of his Moon landing challenge, and so never knew how triumphantly the US had won the space race, relegating the Soviets to the "also ran" slot in technological achievements. Of course, the outcome did not just depend on technologies; it was also a consequence of public support, politics, funding and brilliant management. It brings to mind the possibly apocryphal tale of Queen Victoria, who was told that America had won the first "America's Cup" yacht race. "Who was second?" she is reputed to have asked. "There is no second, Ma'am", replied the embarrassed aide.

The former Texas senator and Vice President Lyndon Johnson was, if anything, even more politically savvy than his former boss. It was no surprise therefore to discover that Houston, Texas, had become the home of the Johnson Space Flight Center (JSC) in 1963, as the space program progressed. It was from JSC that the Moon flights would be controlled, and it was to the Mission Control at JSC that the astronauts walking on the Moon directed their transmissions: "Houston, Tranquility Base here. The Eagle has landed." In Fig 6-6 we see Von Braun again working the presidential route to funding and in Fig 6-7 we note how easily Johnson took over Kennedy's association with the astronauts.

Fig 6-6 Wernher von Braun and President Johnson at Marshall Space Flight Center, March 1968.

(Credit: NASA)

Fig 6-7 President Johnson and Gemini astronauts (Ed White, Gus Grissom, Jim McDivitt) and wives, in 1965.

(Credit: NASA)

The picture shows LBJ with Gemini astronauts Grissom, McDivitt and White with their spouses, soon after the success of the first American space walk, by White, in Gemini IV in June 1965. It was Johnson who, on 16th July 1969, watched the Apollo 11 liftoff (Fig 6-8) that ensured that the country met Kennedy's goals for the space program, thus winning the technological and ideological race with the Soviets. Johnson himself had stepped down from the presidency earlier that year. He therefore watched the launch as it were as Kennedy's surrogate, and as an ex-president. He had, both as vice president (1961-1963), and then as president (1963-1969), been the White House champion during the most dynamic days of the space race. This he had done while dealing with an unpopular war (Vietnam) and domestic turmoil including race riots. He arguably deserves a "Wright Stuff" award almost as much as Kennedy, because he had to do the heavy lifting to push the program to completion, but it was Kennedy who had taken the enormous political risk. And we must be sparing. There may be an even more deserving recipient, with even more direct benefits to the creation of space tourism and the availability of your ticket to space. We shall see.

Fig 6-8 Ex-President Johnson watches the Apollo 11 liftoff, KSC, 16 July 1969.

(Credit: NASA)

In getting to this point of triumph, the country, and therefore the president, had been through the tragedy of the January 1967 Apollo 1 pad fire that took the lives of White and Grissom (to LBJ's right in Fig 6-7) and Roger Chaffee. The risks of spaceflight were apparent to all, and particularly so to politicians and presidents. None of the astronauts on the Apollo missions lost their lives in space, although the Apollo 13 flight was touch and go. Apollo 11 had taken to the Moon a small piece of wood and fabric from the 1903 Wright Flyer to underline the amazing developments in aerospace since the first flight at Kitty Hawk. The returned item can be seen today in a glass case in the Smithsonian in Washington, DC. The Flyer flew for less than two minutes one day in 1903, with 850 feet being its longest flight, yet some of the handiwork of those two brothers from Ohio found its way 240,000 miles to the Moon, and back again. Assisted, of course, by Neil Armstrong from Wapakoneta, Ohio.

The huge irony is that although President Kennedy initiated the race to the Moon, and President Johnson saw it through its tragedies and triumphs; it was President Nixon's name that appears to this day on the plaque on the Lunar Landers, (see Fig 6-9) and Nixon who

was president throughout all the triumphal Moon missions.

Fig 6-9 President Nixon's name is on the Apollo 11 plaque left on the Moon.

(Credit: NASA)

In Fig 6-10 President Nixon is on board the carrier USS Hornet, after the recovery of the Apollo 11 spacecraft and its crew. The crew was held in an isolation chamber for several weeks until medical tests had proven that they had not brought any pathogens back from the Moon. What was particularly cruel for Aldrin and Collins was that Armstrong had arranged in advance for his banjo to be available in the trailer.

Charles Lindbergh was still regarded as a trusted advisor to presidents in the realm of aviation and space at this time. Fig 6-11 captures a moment while he met with President Nixon during the time of Apollo, when the president was considering what space activities could possibly follow the Moon Landings, and in particular what could be achieved within a more reasonable budget than had been spent on Apollo. What would follow would look more like an aircraft and less like a rocket.

Fig 6-10 President Nixon greets Apollo 11 crew (Armstrong, Collins, and Aldrin), in biological isolation on their return from the Moon - USS Hornet, July 24ᵗʰ, 1969.

(Credit: NASA)

Fig 6-11 Charles Lindbergh confers with President Nixon in August, 1972.

(Credit: Lindbergh Picture Collection, Manuscripts and Archives, Yale University Library).

Nixon's presidency was from 1969 to 1974. After Apollo ended in 1972, there was a huge hiatus in space developments. It was as if the nation collectively heaved a huge sigh, said "Been there, done that", and lost all further interest in space. The new NASA Administrator who had to operate during this period (1971-1976) was James Fletcher. The Skylab space station and the joint US-USSR Apollo-Soyuz Test Project (ASTP) mission kept engineers and astronauts employed, during a softening of the

Cold War attitudes. Nixon was president when the decision was made to build the Space Shuttles, but he had left the office in disgrace on August 9ᵗʰ, 1974, following the Watergate scandal, and been replaced by President Ford (1974-1977). The Space Shuttles were supposed to usher in a new phase of spaceflight, with more airline-like operations, and with much reduced costs. It did not, however, work out that way. The Space Shuttles were simply too complex to be able to be recycled quickly and achieve the promised flight rates.

Gerald Ford was president when the Smithsonian Institution's new National Air and Space Museum (NASM) was opened on the National Mall in July 1976 in Washington DC (Fig 6-12). We would have to wait until April 1981, and another president, for the first flight of the Space Shuttle. Meanwhile, President Ford awarded yet another medal to Chuck Yeager in the by now well-established presidential tradition.

Fig 6-12 President Ford and Mike Collins at the opening of the new National Air and Space Museum, Washington DC, July 1ˢᵗ 1976.

(Credit: National Air and Space Museum)

The new NASM museum's first Director was Mike Collins, who had been appointed relatively fresh from his journey to the Moon with Apollo 11. The Apollo 11 capsule took center stage in the main hall of the new museum, which quickly became a major attraction in DC. People who were not alive during the Apollo era can nowadays learn about the experience through its exhibits. It is the main place where the public can get to know about space in the US. Around 5,000,000 visitors per year, from all nationalities, go to see its exhibits. It now contains exhibits devoted to space tourism, too.

The Carter presidency (1977-1981) was not noted for any space initiatives, and so we see that Fig 6-13 is a commemoration of past glories with the tenth anniversary of the first Moon landing with the Apollo 11 crew in 1979. It was Carter's misfortune to be president during the gap following the last Apollo flight (which was for the Apollo Soyuz Test Project (ASTP) Mission in 1975) and before the first Space Shuttle flight in 1981.

Fig 6-13 Apollo 11 crew (Mike Collins, Buzz Aldrin, Neil Armstrong) with President Carter on July 20th, 1979.

(Credit: Whitehouse.gov)

President Ronald Reagan (1981-1989) watched over the successes of the early Shuttle flights as the Soviet Union collapsed. But he also had to deal with the aftermath of the first Shuttle disaster (Shuttle "Challenger" was destroyed by an explosion soon after liftoff in January 1986, killing all onboard. The explosion was due to a failure of a joint between segments in one of the Shuttle's strap-on solid rocket motors). In Fig 6-14 we see the former California governor with his wife Nancy out at Edwards Air Force Base to greet the arrival of one of the early Shuttle flights (it was Shuttle STS-4 in July 1982). Edwards continued to provide service to the space program as an emergency landing facility for the Shuttle. In 1986, President Reagan awarded Medals of Honor for the first non-stop round the world flight, without refueling, of Dick Rutan and Jeana Yeager, whom we saw in Chapter 3. And for good measure, the president awarded yet another medal to Chuck Yeager (Fig 6-15).

Fig 6-14 President and Mrs. Reagan greet crew (T.K. Mattingly and H.W. Hartsfield) of Space Shuttle "Columbia" after landing at Edwards in July 1982.

(Credit: Whitehouse.gov)

Fig 6-15 President Reagan awards medal to General Chuck Yeager.

(Credit: Whitehouse.gov)

The first President Bush (1989-1993) tried to re-invigorate the country's lapsing interest in the Moon and space, with his Space Exploration Initiative (SEI) of 1989. He was to find that the public and Congress were not interested, and no funds could be allocated at a time when the country was entering the Gulf War. Fig 6-16 establishes the Apollo 11 crew in October 1989, now 20 years after their Moon landing, yet again being called to serve to support a president's space initiatives.

Fig 6-16 President G.H.W. Bush and Apollo 11 crew (Collins 2nd from left, Armstrong behind president, Aldrin at right), in October 1989.

(Credit: Whitehouse.gov)

Fig 6-17 President Clinton makes award to astronaut Jim Lovell, July 1995.

(Credit: Whitehouse.gov)

President Clinton's era (1993-2001) brought no significant new space ideas, except the slow start to building the International Space Station. Although by this time there were the stirrings of an informal "new space" movement, whose members were dissatisfied with the progress of NASA since Apollo, and who therefore wanted to create a commercial alternative. This movement would come to embrace the idea of space tourism. Meanwhile, Clinton's NASA Administrator Dan Goldin was not winning much support with his "Better, Faster, Cheaper" mantra. Most experienced engineering managers thought that you might be able to get any two of the three goals at any one time, at the expense of the third, but you could not get all three simultaneously. Jim Lovell's book about his Apollo 13 near disaster came out in 1994, and was made into a successful movie in 1995, renewing public interest in the achievements of the 'sixties, and President Clinton awarded the Congressional Space Medal of Honor to the popular ex-astronaut (Fig 6-17). He had already received the Medal of Freedom from President Nixon in 1970. During the Clinton presidency, there were unsuccessful attempts to design a single-stage-to-orbit vehicle. One of these was the DC-X (1993-1996), which Apollo astronaut and Moonwalker Pete Conrad helped design and fly as a commercial space initiative. More about this later.

We have followed the arc of presidential involvement with space matters, and so far awarded only one "Wright Stuff" accolade in this category. President George W Bush tried again to interest Congress and the American public in space in 2004 through his Vision for Space Exploration (VSE), which embraced a return to the Moon and eventual onward journey to Mars. He established various Commissions, both before and after his decision, and Apollo 11 astronaut Buzz Aldrin was a Commissioner on one of them, arguing for consideration of space tourism. Buzz's contribution has been substantial to the emergence of space tourism, and he has used his fame and access to keep pushing forward for the achievement of that goal, but it will be with the second President Bush that the effort will be seen to bear fruit. There was not intended to be extra funding provided for achieving the Vision; it merely signified a change of direction, and a reallocation of existing NASA funds. There had been a second Shuttle disaster under his watch. The shuttle "Columbia" broke up on re-entry in February 2003 due to a hole in the thermal protection of a wing leading edge caused by debris falling from the external fuel tank during launch. Bush listened to the advice to phase out the aging Shuttle's missions, and to replace the vehicle with something new, suited to Moon and Mars journeys. Work began on the Constellation program. The new Administration of President Obama, however, taking office in 2009, has decided not to continue with the VSE as originally conceived, or at least not to proceed with the Constellation series of spacecraft and launchers. And in any case it is not for the VSE that this author awards the second Presidential "Wright Stuff" accolade to President George W Bush, the 43rd President of the USA. It is for something else, which was very risky for a politician and a President to do. And you, as an upcoming space tourist, need to thank him for this.

President George Bush 43 signed into law a regime for space tourism that fully accepted that the venture is risky, while understanding its business potential. Even though as president he had observed the nation's mourning over the accident to the "Columbia" and the loss of its astronaut crew, he nevertheless recognized that space tourism reflected some essential American values. It is risky, it pushes back frontiers, it's fun, and you can make a buck doing it. At the time of writing, it seems unclear whether any other national government would put such a regime in place to enable the creation of a space tourism industry. It is largely about freedom. The regulations make it clear that it is *not* the government's business what you or anybody wants to do with their time and money in space, (so long as it is legal and can be taxed!). That includes the freedom to do very risky, even life-threatening, things. It *is* however the government's responsibility, the regulations make clear, to make sure that nobody in the *uninvolved* public should be put at risk by your flight.

It is entirely possible that there will be accidents in the early years of space tourism, but this president took that into account in reaching his decision to sign the Commercial Space

Launch Amendments Act (CSLAA) into effect (December 2004). Thus he earns the second Presidential "Wright Stuff" accolade for moving this nation forward towards its next step in space, the space tourism revolution. He knew that space tourism brings potential benefits in a number of ways, including employment, new businesses, transformative technology, improved safety, lower costs for payloads into space, and a more airline-like way of operating. Bush 43 followed the developments of the X-Prize, and even put in a congratulatory telephone call from the White House in October 2004 to Burt Rutan's team on his successful completion of the Ansari X-Prize competition. Fig 6-18 has the Apollo 11 team again doing their service in the Oval Office, when Buzz Aldrin's efforts on behalf of space tourism have at last found their responsive president. If we had three awards to offer in this category, Johnson would also have received one, but two awards are probably more than enough to be given to politicians.

Fig 6-18 The Apollo 11 crew (Collins, Armstrong, Aldrin) in the Oval Office of the White House with President George W Bush, July, 2004. This was the 35th anniversary of their Moon landing mission. President Bush signed the enabling legislation for the creation of the space tourism industry.

(Credit: Whitehouse.gov)

As if to demonstrate that all presidents share a collective need for astronaut photo opportunities, Fig 6-19 represents the beginning of a new Administration, with the Apollo 11 crew once again turning up at the White House to meet the

president. This time, it was the 40th anniversary of their moon-landing mission, on July 20th, 2009. In February 2010, President Obama offered a new direction for NASA, one which involves using commercial launch and spacecraft providers to get crews to low Earth orbit, and which provides major investments in fundamental space technologies which would be required before undertaking long-distance interplanetary crewed missions. In 2010 the FAA created a Center of Excellence (COE) for Commercial Space Transportation, and a Commercial Space Committee was added to the NASA Advisory Council's toolkit, to help guide the agency into the new era. The odd aspect to this development is that a Democratic President has introduced a very Republican concept – the idea of NASA not building and flying its own spacecraft, but using private industry to provide a taxi (or maybe rental car) service, for a fee, to get NASA astronauts into space. At the time of writing (March 2010) Congress had still not given its verdict on the new plan. If it goes ahead, however, it will provide a boost for the space tourism concept.

We have checked out the record of a procession of US presidents in assessing whether they have helped you to obtain your ticket to space. Fortunately for you, we were able to find two who took the necessary steps during their terms of office to take a deep breath and sign the enabling legislation. Kennedy's Moon program gave us all of the necessary experience in human spaceflight, including maneuvering, rendezvous and docking. Bush 43 provided the incentives for a commercial industry to provide public access to space. But is the general public ready yet to follow this presidential leadership? In the era following the events of the terrorist attacks of 9-11, has the collective US public consciousness become too risk-averse to endorse the necessary next steps? Is this still "the home of the brave"? Maybe the community of artists can be counted upon to raise the public awareness, and do their part to make the new business of space tourism a reality. They have helped in the past, as we shall see in the next chapter.

Fig 6-19 The Apollo 11 crew meet with President Obama on the 40th anniversary of the first Lunar Landing, July 20th, 2009. The President would later announce a new era where NASA would rely on commercial launch and spacecraft providers to take NASA astronauts into low Earth orbit. Maybe orbital space tourists could occupy any spare seats on these new commercial spacecraft.

(Credit: Whitehouse.gov)

John F. Kennedy

George W. Bush

5 - ARTISTS

"Someday, humankind will populate the cosmos. The wonderful possibilities are endless – as endless as the infinite regions that surround Planet Earth. The human spirit, driven as it is with an insatiable desire to know, to explore, and to understand, will continue forever to reach upward and outward."

Artist Robert McCall, 1992
(from " The Art of Robert McCall – A Celebration of our Future in Space"- Ref 6)

There have been aviation artists ever since there has been aviation. Fig 7-1 provides a typical early poster designed to capture the public's attention about the joys of flying. We notice that they tended to underplay some of the realities of the experience. The later ones about passenger flying (Ref 1) did not perhaps give a full rendition of the courage required, or the vibration, noise, cold, smells, grease, turbulence and earache suffered on those first flights. But they certainly captured the public's imagination.

There were aviation artists who documented the feats of airmanship in the two World Wars, provided marketing images for manufacturers of each new aircraft arriving on the scene, and evoked excitement for the public about the possibilities that the new airlines were offering. Trips to exotic places were on offer. Lindbergh has himself been celebrated in countless statues and paintings, and the huge mural at San Diego airport in California (Fig 7-2) is an example.

Fig 7-2 The mural of Charles Lindbergh at San Diego International Airport (photo in 1990's).

(Credit: Author)

Fig 7-1 Early aviation art poster – artist Charles L Brosse (1910).

(Credit: Allposters.com)

Following on from, and running parallel to, aviation art is space art. Both aviation and space art represent a new genre, unknown and unacknowledged to many more traditional artists. It is dismissed by some as "mere" illustration. It has nevertheless been very important in influencing public opinion in favor of space travel from its earliest times. Certainly, the early NASA administrators, such as James Webb, were aware of its importance, and put in place various schemes to make sure that the activities at

NASA were captured not only photographically but through the eyes and paintbrushes of artists. The NASA Art Program was established in 1963. Art can capture things that photographs cannot. Sketching of concepts is also an important step for an engineer/designer to visualize the finished device or craft, and for our purposes we also include this under the rubric of "art".

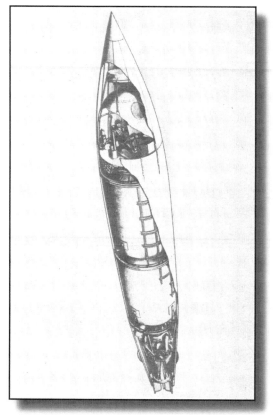

Fig 7-3 RA Smith line drawing design for a sub-orbital manned vehicle (1946). Fifteen years later, Alan Shepard would rocket into space in a very similar vehicle configuration (See Fig 4-34 for comparison).

(Credit: British Interplanetary Society/www.BIS-spaceflight.com)

Art can present us with visions of places where it would be impossible to go, and where it would even be impossible for robots to go: for example, positioned directly above a Moon rocket as it lifts off heading for space, or in a galaxy hundreds of light years away. And space art does not have to be representational. It can capture moods and feelings, and the challenge and the energy needed to go, and the sense of a nation working to meet that challenge. It can tell us what

it will be like to be there, when we get the chance to go as space tourists. Art, in the fullest sense of the word, can include movies and music too. It is an important element in science fiction. Sales of the music recording of "*Also Sprach Zarathustra*" by Richard Strauss certainly benefited after the release of the movie "2001: A Space Odyssey", where it was a peaceful, hopeful and ethereal background theme. You yourself have probably been influenced by artists' renderings of your future space trip experience into paying for your ticket to space.

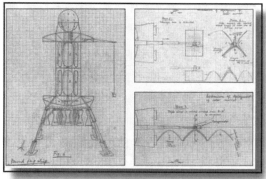

Fig 7-4 von Braun pencil drawings of vehicle concepts on his favorite Keuffel and Esser engineering paper (1952).

(Credit: Philosophy of Science Portal)

We have already seen some space art in this book; as the author said at the outset, the several themes are intertwined across the chapters. So, in Chapter 4, Fig 4-24 referred us to some of the work of the BIS artist Ralph A. Smith, who died in 1959 -even before the first manned space flights- but nevertheless managed to capture so much of what was about to take place (see Ref 2). Smith had been an architectural decorator and design draughtsman before the Second World War, and he not only produced the BIS artwork, but labored over its basic engineering content also. Fig 7-3 indicates another of his designs, which in 1946 foresaw Alan Shepard's 1961 Mercury "Freedom 7" Redstone ride into sub-orbital space: the ride that has so much in common with your modern sub-orbital space tourism trip. People have been waiting an awfully long time to enjoy the trip you are about to experience.

Wernher von Braun himself used to draw his concepts on squared engineering paper to work out relativities between various components and envision the scale of the finished vehicle (see Ref 3). We see some of his drawings in Fig 7-4, which include his 1952 design for a lunar cargo ship.

Von Braun was, however, acutely aware of the importance of public support for his proposed space ventures (see Ref 4). He had been brought to the US after the Second World War and hoped to continue his ambitions regarding space flight, and he knew that in his new environment, and new country, public opinion needed to be harnessed in order that funding would be made available. If the most important rocket fuel was dollar notes – "no bucks: no Buck Rogers", then it would be the general public who ultimately would have to authorize the funds. We have already seen how von Braun worked at the political level to move forward with his ideas. We now look at how he addressed the need for public support. He wrote a series of books, and some associated articles for "Collier's" magazine, and called upon the services of the artist Chesley Bonestell to convert his designs into believable images. Fig 7-5 places Bonestell with von Braun, in front of a model based on the von Braun lunar cargo ship sketch in Fig 7-4.

Fig 7-5 Artist Chesley Bonestell with von Braun (around 1952) and a model of a lunar cargo ship.
(Credit: Collectair.com)

Name Chesley K Bonestell

(Credit: NASA/New Mexico Museum of Space History)

Summary Description
Legendary space artist who captured von Braun's visions

Date of Birth
January 1 1888, San Francisco, California, USA

Date of Death
June 11 1986, Carmel, California, USA

Nationality US

Achievements
Inspired generations of space enthusiasts
Received BIS Bronze Medal

Specific help for Your Ticket to Space
If a picture is deemed to be worth a thousand words, then Bonestell's contributions to von Braun's engineering concepts were worth many manuscripts of engineering notes. He brought the dream of space to an Earth-based reality. His paintings show the details of extra-terrestrial soils , and vehicle tracks, in such a way that the public for the first time could really believe it was possible, and in fact almost that it had been done! He illustrated von Braun's most famous books and worked with the Disney organization as they produced the TV movies and the Tomorrowland Experience. The US public was therefore ready to support the effort during the sixties to open up the world of space. He even did some work on point-to-point sub-orbital space transportation in the same period, and so he still remains ahead of the actual achievements of the space industry. Maybe space tourism will develop in that way, so that future space tourists will experience the Bonestell ride from LA to Sydney in only an hour.

Fig 7-6 illustrates how Bonestell brought life to the design through his "Collier's" artwork. The "Collier's" articles appeared through the period from March 1952 to April 1954. In effect, they prepared the American public for the arrival of the space age, which began in 1957 with Sputnik 1. The articles had a huge impact on the American public. This led to von Braun's next venture using art to capture the public imagination. Bonestell became the most universally recognized space artist for his contributions. He was able to accurately evoke the view of Earth from space even before anyone had actually seen it. Bonestell died in 1986, aged 98, having seen some of his artistic projections come to life. However, his magnificent series on human flight to Mars, done at the same time as his lunar work with von Braun, will have to wait a few more decades before being evaluated against the test of reality.

Fig 7-6 Chesley Bonestell cover illustration for "Collier's" magazine, Oct 18th, 1952.

(Credit: JSC.NASA.Gov)

Von Braun began to work with Walt Disney in the early '50s (note Fig 7-7, which includes von Braun, Disney, Willy Ley and Heinz Haber) to produce a television series of three films, "Man in Space", "Man and the Moon", and "Mars and Beyond", based upon his concepts. Ley was organizer of the Hayden space symposium, and Haber was a space medicine expert. They had all contributed to the "Collier's" articles. Willy Ley

had been a founding member of Germany's rocket society VfR in 1927, and wrote several early spaceflight books (see Ref 5). He emigrated to the US before World War II and later worked with Bonestell and von Braun. He died just one month before the Apollo 11 Moon landing, but has a far side crater named in his honor.

The next image of box-top art, Fig 7-8, testifies to Ley and his work of publicizing the space program. The author was an avid modeler as a teenager in the UK, and Willy Ley's kits (on those rare occasions when they could be obtained from the US) always had a special ring of authority to them. Ley's book (Ref 5) was a great source of rocket data, and the author has a prized 1958 edition in his collection.

Fig 7-7 Wernher von Braun, Willy Ley, Walt Disney and Heinz Haber in 1954 discussing the "Man in Space" film for the "Disneyland" TV show, with a model of a two-stage rocket consisting of a V2 and a WAC Corporal upper stage.

(Credit: NASA)

The first Disney film, "Man in Space", aired in March 1955, and was seen by *an estimated 100 million viewers.* Apparently President Eisenhower was impressed by the series and obtained a copy of it. It was this series that helped prepare the American public for the space age, and made von Braun a publicly recognizable figure. The third and last film in the series was aired in December 1957, just after the Soviet Union had launched its first artificial satellite. Disney lavished what in that era was an unprecedented sum of $1M on the production of the three space films. There were comic book versions, too, with a wide distribution, and the author was able to enjoy reading them as a schoolboy in England at the time.

Fig 7-8 Willy Ley's space publicity machine attracted the minds of the young. A 1958 space-themed plastic model construction kit, endorsed by Willy Ley.

(Credit: Monogram)

Fig 7-9 Norman Rockwell at Glen Canyon Dam, where he was doing his signature careful preparatory work for one of his classic paintings, which had a Southwest theme. Rockwell used to work a great deal from photographs, and he visited all the locations whenever he could. In the case of the Moon Landing, however, he could not be at the Sea of Tranquility, but he visited the Apollo 11 astronauts when they were training to take their first steps on the Moon.

(Credit: USBR.gov)

Space as a subject has attracted many artists and illustrators, and their work has often found itself on murals of space museums, in articles about space in popular magazines, or even in the designs of postage stamps. Fig 7-10, for example, is a rendering by the popular artist Norman Rockwell (Fig 7-9) of the first Moon landing by the crew of Apollo 11, with Neil Armstrong taking his first step on the new world. The image carries more drama than the photos brought back by the crew from the Moon's surface, and in fact is not a strictly accurate rendering of the lunar surface, or of the heavens seen from the Moon. However, it is undeniable that the artwork evokes a strong emotion that captured the nation's mood on the occasion of the Moon landings.

Fig 7-10 Norman Rockwell's 1969 painting captures the drama of the first lunar landing.

(Credit: NASA.gov)

Name Robert T. McCall

(Credit: Author)

Summary Description
Space artist, muralist and postage stamp designer

Date of Birth
December 1919, Columbus, Ohio, USA

Date of Death
26 February 2010, Scottsdale, Arizona, USA

Nationality US

Achievements
Produced massive space-themed murals at NASA centers and US public space museums.
Recorded early astronautical history via his artworks.

Specific help for Your Ticket to Space
Bob McCall is in no danger of being forgotten any time soon, because his major artwork murals celebrating man's conquest of air and space are in many galleries and museums, at Edwards Air Force Base, the Johnson Spaceflight Center, the Pentagon, the Air Force Academy, the Smithsonian, and elsewhere. His work conveys a wonderful optimism about mankind's future in a world where the solar system has become a natural part of our living space. To McCall, space tourism is an inevitable first step for us to begin to understand our environment, but only a small first step; it is not Neil Armstrong's "Giant Leap", but only the beginning of mankind's space adventure. Are you ready to take your seat, and thus help achieve McCall's wider vision?

Rockwell used to provide illustrations and cover art for the very popular "Saturday Evening Post" magazine, and Fig 7-11 displays an image he created for the August 1969 special commemorative edition of "LOOK" magazine. It captured, in a triple-page foldout spread, the workforce behind the Apollo 11 astronauts who had just completed their historic first Moon landing.

Fig 7-11 Norman Rockwell's commemorative painting of the workforce supporting the Apollo 11 astronauts, 1969. Amongst those depicted are: the backup crew, the insertion technician Joe Schmitt, Chuck Berry - the astronauts' surgeon, Al Shepard, George Low, Wernher von Braun, Kurt Debus, George Mueller, Robert Gilruth, Chris Kraft, the astronauts' wives and various range staff. Taking a keen interest in von Braun as both a space traveler and artist is Alexei Leonov.
(Credit:NASA via Ed Hengeveld)

Another significant space artist has been Bob McCall, and his giant mural "A Cosmic View", depicting the Apollo missions, occupies a whole wall of the entrance lobby at the National Air and Space Museum in the Mall at Washington DC. This one image is viewed annually by 5,000,000 visitors to the museum (see Fig 7-12 and Ref 6). Author Isaac Asimov once described McCall as "The nearest thing to an artist in residence from outer space". The author met McCall in San Diego in 1995 at a conference honoring the Apollo astronauts. McCall demonstrated his immediate sketching facility on the spot by means of a 10-second glorious solar system perspective drawn directly onto the event program, which captured a wonderful optimism about man's place in the universe. So the author now owns an original McCall!

Fig 7-12 Bob McCall's "A Cosmic View" mural (146' x 46', acrylic on canvas, 1975) at the National Air and Space Museum, Washington, DC. The mural is L-shaped, and at the far right end, the inspiring work is six stories high.

(Credit: Author)

In Fig 7-13 we see the base of McCall's 1975 NASM mural "A Cosmic View" being used as a backdrop for 5 of the lunar astronauts (4 of whom walked on its surface) at the time of the Centenary Celebrations for the Wright Brothers. Insiders at the National Air and Space Museum can point out where both the artist's wife, and the Moonwalker Alan Bean, made symbolic cameo contributions to the artwork. Bean's contribution is a tiny rendition of the astronaut badge that he threw away at his Apollo 12 landing site. Both the Wright Flyer and the Apollo 11 capsule were prominent during the November 2003 events. It had taken McCall eight months to create the mural that in one section is six-stories high.

Fig 7-13. Moon travelers with Bob McCall's "A Cosmic View" mural as backdrop. National Air and Space Museum, December 17th, 2003 (Buzz Aldrin, Alan Bean – who had made a minor symbolic addition to the artwork, Dick Gordon, Gene Cernan, Charlie Duke). This was one hundred years to the day after the first flight of the Wright Brothers.

(Credit: Author)

The next figure (Fig 7-14), also presents Bob McCall's work, where he captures 100 years of combined aviation and space history. The original 2003 work is at NASA's Dryden research facility at Edwards Air Force Base, California, and the full scale replica is at the National Air and Space Museum's Udvar-Hazy Center, near Washington's Dulles Airport. He also has murals at NASA's Johnson Space Center in Houston, Texas and at the Kansas Cosmosphere in Hutchison.

Fig 7-14 Bob McCall's full scale replica of the NASA Dryden mural at the Udvar Hazy Center of the NASM in Virginia.

(Credit: Author)

Bob McCall's work spanned the very large and very small, and often found its way onto US space postage stamps. He provided the artwork for 21 different space-themed US postage stamps (e.g. Fig 7-15). He also designed astronaut mission insignia, such as the Apollo 17 patch, and patches for various Space Shuttle crews.

We already saw (Fig 4-25) the majestic image that McCall created for Stanley Kubrick's seminal 1968 science fiction film (of Arthur C Clarke's book of the same name) "2001: A Space Odyssey".

Bob McCall died in February of this year (2010), and we can safely say that his artwork will continue to inspire for generations to come. His comments make his legacy very important and optimistic: "I think when we finally are living in space, as people will be doing soon, we'll recognize a whole new freedom and ease of life. These space habitats will be more beautiful because we will plan and condition that beauty to suit our needs. I see a future that is very bright."

Fig 7-15 Bob McCall designed US postage stamps in 1992 showing US and Russian space achievements to that date, from Sputnik 1 and Vostok, through Mercury, Gemini, the Moon landings, ASTP, the Space Shuttle and Space Stations.

(Credit: US Postal Service. Space: block of four stamps © 1992. All Rights Reserved. Used with Permission)

In 1997, a major space art mural was installed in El Segundo, CA, a suburb of Los Angeles adjoining the Los Angeles International Airport. The area had long been associated with aviation and space developments, and was particularly related to the work of Howard Hughes, both as an aircraft manufacturer, aviator, and subsequently as a manufacturer of communications satellites. Fig 7-16 shows this artwork, commissioned by Hughes Electronics Corporation, which is 118 feet wide and three stories high, and adjoins a parking lot. The artist was Scott Bloomfield, of Long Beach, California, and he managed to capture the major aerospace developments from the Wright Brothers all the way to communications satellites and Space Shuttle flights in his work. Howard Hughes himself is prominent at the left edge of the mural, and the record breaking Amelia Earhart (who ensured that women were accepted as equals in aviation, when she flew the Atlantic solo just 5 years after Lindbergh) looks rather poignantly from the center. Bloomfield, who paints abstracts and landscapes as well as portraits, himself described the mural as "…a view of Heaven. It's people who have no limitation in scale..." Scott Bloomfield, though not strictly a space artist, has made a great contribution in this otherwise uninspiring parking lot.

Fig 7-16 The January 1997 mural "Spirit of Aerospace" by Scott Bloomfield in a parking lot, 520 Main Street, El Segundo, California.

(Credit: Author)

For a special insight into the minds of space travelers, we are fortunate that two former astronauts are themselves space artists, and therefore have a unique viewpoint from which to convey their images of space exploration, and thereby convey their joy and enthusiasm to the general public.

Fig 7-17 shows the Russian cosmonaut Alexei Arkhipovich Leonov with a book of his space art. He commanded the Soviet part of the ASTP Mission. He flew twice in the Soviet era. On his first flight, in Voskhod in 1965, he became the first person to conduct a spacewalk. In later years, as head of the Russian astronaut training center, Leonov created an encouraging environment for the first space tourists as they began their tests in Moscow's Star City (see Ref 11). The author met Leonov in 2001 in San Diego when Leonov was being invested in the International Aerospace Hall of Fame at the Balboa Park Aerospace Museum. He has a fine sense of humor and was an excellent ambassador for the former USSR when he was chosen to lead the Russian part of the joint US/USSR ASTP project. When the two spacecraft had rendezvoused and docked the two crews got together for a meal of astronaut food, which back in 1975 was not very palatable. A press conference took place with questions coming

in Russian and English from journalists in the mission control room. When asked about the food, Leonov answered, in perfect English, "The best thing about any meal is not what you eat, but with whom you eat!" We observed his keen interest in the work of another space artist in Fig 7-11.

Fig 7-17 Cosmonaut artist, author, and space tourism facilitator, Alexei Leonov.

(Credit: David M Scott/Apolloartifacts.com)

Through Fig 7-18 we see how Leonov portrayed that first spacewalk. The Voskhod spacecraft was really just a one-man Vostok, with seats for two and an exit airlock arrangement squeezed in. The airlock was made of inflatable materials and can be seen jutting out and upwards from the side of the spacecraft in Leonov's painting. His life-sustaining umbilical cord can be seen snaking its way back into the airlock. This first spacewalk nearly proved fatal. Once he was outside free floating, Leonov's spacesuit began to expand in the vacuum of space, and he could not initially re-enter the airlock. He was in big trouble. Imagine the terror at being trapped outside of your spacecraft in orbit. He had to take the risky decision to let some of his precious air escape so that the suit would deflate and he could return to the safety of the capsule, which he did before almost collapsing with exhaustion. Of course, Leonov's painting focuses on the romance of his spacewalk, the vistas, the curved horizon, the black sky and the stars seen in daylight, the cloud patterns below. He even took colored pencils into orbit so that he could faithfully render the colors that he observed, especially those at the Earth's horizon at the sixteen times a day when there is a sunset or sunrise in orbit. This is the view you will shortly see on your space tourism flight: black sky, even in daylight, curved horizon, thin film of the atmosphere, and cloud patterns way down below.

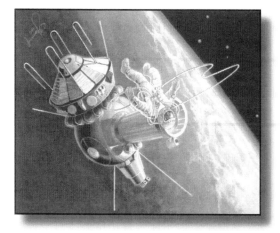

Fig 7-18 Leonov's depiction of the first spacewalk, 18th March, 1965.

(Credit: Pacific Coast Galaxy)

We mentioned the American Apollo astronaut Alan Bean, who flew to the Moon on Apollo 12 in November 1969 for the second lunar landing, in connection with the McCall mural at the National Air and Space Museum in Washington DC. Bean has devoted his life since returning to Earth to painting images inspired by his experiences walking on the Moon (see Ref 10). "It's beautiful out there!" he had declared as soon as he landed on the Moon on the Ocean of Storms. Like Leonov, Bean also has stayed away in his painting from underlining the courage that was needed in those early missions. One of the most difficult questions to answer for those who walked on the Moon was apparently "What color was the Moon?" This difficulty was because the impression varied dramatically with sun angle. Bean did a great deal of experimentation to try to convey this fact of Moonwalking. He has an interesting technique, and uses some of his Moonwalking and lunar geology tools to add texture to his work in its early stages. Fig 7-19 shows Bean in his studio. Fig 7-20 shows a visitor enjoying Bean's major 2009 Exhibition at the NASM in Washington DC, which coincided with the 40 years commemoration of the Apollo lunar landings.

Since Bean's artistic output all concentrates on the Moon, we may say that his artwork has not directly helped space tourism, but he certainly works hard to capture the thrill of spaceflight. He is the only artist amongst the twelve men who walked on the Moon, and feels a certain obligation to convey that experience to the best of his ability. Al Bean is in addition a very effective motivational speaker to young people, who we all realize are the space tourists of the future. The author watched him effortlessly charm a group of children, who had come out to hear about his space experiences, in Tucson, Arizona in 1998. Bean was subsequently present at the Wright Brothers' Centenary of Flight Conference in Washington DC, December 18, 2003. Afterwards, in discussions about space tourism, he indicated his general support: "It's very important for the young people to be able to have grand dreams and be challenged to fulfill them. If space tourism can be part of that, then it's good."

Fig 7-19 Alan Bean in his art studio, October 2008.

(Credit: Russo/National Air & Space Museum)

Mike Collins, who was the Command Module Pilot on the Apollo 11 mission in July 1969, has in later life also become a hobbyist water color painter. So far his work has focused on the beauty of the Earth (often at his favorite fishing spots), and not the arid surfaces of the Moon or the vacuum of space. Leonov and Bean have generally devoted their paintings to studies of the past. The early artists, such as RA Smith and Chesley Bonestell looked to the distant future in producing their iconic images.

Who is today producing imagery that can capture the imaginations of today's public, to get their support for the nation's next steps into space? And who is doing artwork that brings the idea of space tourism to the fore? The answer is probably that there are not enough of them, but some have been carrying the flame. We can list a number of space art practitioners, and show some examples.

Fig 7-20 Exhibition of Alan Bean's artwork at the Smithsonian's National Air and Space Museum, 2009.

(Credit: Author)

Active space artists today include Pat Rawlings, David Hardy, Paul Calle, Chris Butler, Kim Poor, Ed Hengeveld, Pamela Lee, Phil Smith, Don Davis, William Hartmann, Andrei Sokolov, and many others, and their original and lithographic artwork is in demand (see Ref 7, Ref 8, Ref 9, Ref 12, Ref 13). A good web-site where this work can be accessed is at www.IAAA.org. In the following pages it has only been possible to offer a glimpse of some of this work, and it is not even necessarily representative of the full panoply of space artists or the genre as a whole. The author has tended to include images that are most connected in some way to the space tourism business.

Pat Rawlings (Fig 7-21) has been an active space artist for decades, on the staff of the consulting firm SAIC, and helping NASA to envisage its various engineering concepts, but also addressing such future space tourism concepts as a Lunar Olympic Games. Can you imagine what it would be like to go pole-vaulting in a 1/6 gravity environment, or to go there as a tourist to watch? Rawlings can. Already, at the time of writing this book, there are planned space tourism trips to Lunar orbit being offered by the company Space Adventures (about which we will learn more later). Rawlings worked with Peter Diamandis, a key player in the space tourism movement, and one of the International Space University (ISU)'s founders, when it was created. Fig 7-22, which was painted in 1987, provides a whimsical look at the ISU at its

eventual orbiting space campus, which has a certain architectural affinity with MIT, Diamandis's *alma mater*. The author has been a visiting lecturer at the ISU's summer schools, which have taken place at campuses all around the world, but thus far they have not taken place in Earth orbit!

Fig 7-21 Pat Rawlings with his 2004 painting of Neil Armstrong against a backdrop of the Moon cleverly substituting for a space helmet.

(Credit: Roger Weiss)

Fig 7-22 Pat Rawlings' 1987 image evokes an International Space University campus in orbit. Alumni from the ISU are now in leadership positions around the world in space program developments.

(Credit: International Space University)

Fig 7-23 Pat Rawlings' image of the X-Prize competition, conceived in 1998. For comparison, see Fig 2-1 for the reality of September 2004.

In 1998, Pat Rawlings also produced the definitive image of the X-Prize competition and the X-Prize Cup events (Fig 7-23) which, as we have learned, have led directly to the creation of the sub-orbital space tourism sector. The Ansari X-Prize Cup would eventually take place in 2004 at Mojave Spaceport, California. It turns out that, as we shall learn in the next chapter, the 1998 Rawlings painting was a pretty good rendering of the subsequent 2004 events. 20,000 people turned out before sunrise to witness the start of a new phase of the grand space adventure. There were several full scale mockups of competing craft, but only one fully operating prototype, the SpaceShipOne, which the assembled crowd would watch head off into space. The carnival atmosphere captured by Rawlings was real enough. Folks were selling hot dogs and binoculars, and picnicking, while young children and family pets joined in the general festivities.

As a space artist, Phil Smith really understands the subjects of his artwork, because in his "day job" he is an analyst and futurist working in various technology consulting and forecasting operations.

He was, for example, one of the forecasting team at Futron Corporation when that organization was doing long range forecasting work to study the potential commercial and government payload markets for the successor to the Space Shuttle. Phil has produced space art in several styles, and using several different media. His images are generally representational, but he has a vision that brings the distant future within our grasp. Fig 7-24 introduces a Phil Smith cover for an early study into the potential market for space tourism. In his illustration, Smith conveys a feeling of a distant era when point-to-point space transportation might take place from spaceports. The business executives in the holding lounge awaiting their next flight are perhaps going from Los Angeles, California to Sydney Australia, with an expected flight time of 45 minutes. We shall see later in the book that such ideas are still many decades away from fulfillment, but Smith challenges us to consider the normality of such a venture. Point-to-point space transportation is a category which may prove to be more difficult to achieve than either orbital or sub-orbital space tourism. It might emerge as a cargo service rather than a passenger service, but artists have an important role in allowing the rest of us to consider the ramifications of future concepts.

Fig 7-24 Phil Smith's cover art for the Futron/ Zogby market study captures the promise of the future of space tourism (colored pencil on board, 2002).

(Credit: Philsmith.us)

Fig 7-25, also by Phil Smith, reveals the scene over 60 miles above the Mojave Desert in California in 2004 when SpaceShipOne reached its highest point, thus opening up the era of suborbital space tourism. The same desert backdrop has been present for all the pilots of the X-planes flying from Edwards Air Force Base since the 1940's. As the flight envelope reached ever higher, the pilots would see more of the curved horizon, and the sky would become darker, until the craft entered into the space environment. The SpaceShipOne test pilot (either Brian Binnie or Mike Melvill) in Smith's painting had entered a very quiet regime when the rocket motor had ceased from firing. The pilot experienced several minutes of weightlessness under a totally black sky while in brilliant sunlight, before re-entering the atmosphere and returning to Mojave spaceport. On return, the environment was very different. Thousands of enthusiastic people had assembled to loudly applaud the achievement.

Fig 7-25 Phil Smith's rendering of the first civilian spacecraft SpaceShipOne over the Mojave Desert, 2004 (Acrylic on Masonite).

(Credit: Philsmith.us)

Another working space artist, who has helped to bring the world of space tourism into reality, is Mark Maxwell, whom we see in Fig 7-26. In it is some of his artwork that captured the experience being offered by the startup orbital space tourism company Transformational Space, Inc.

Fig 7-26 Mark Maxwell with his artwork depicting the Transformational Space Corporation's CXV orbital space tourism vehicle (2005).

(Credit: Author)

Even though it could be argued that artists rarely have to be concerned about risk to the degree of our other categories, nevertheless they have played an important part in getting you a ticket to space, so an award for the category must be made. So, who gets the "Wright Stuff" award for space art? Although Bob McCall and Pat Rawlings could each make a good claim, probably no-one would disagree with the nomination of Chesley Bonestell as the most worthy recipient, for his far-sighted portrayals which helped make the future believable.

Artists have throughout the ages been counted upon to reveal and celebrate the best of the human spirit, and they certainly did so during the Apollo era. At present, we look to their collective sensibilities to recognize and record how the space tourists' perspective reveals the beauty, fragility, and essential uniqueness of the home planet. We shall keep our second award for the time being, until an artist emerges who can grasp the public imagination in as far reaching a way as Bonestell, and can thereby help the next steps into space, including space tourism, to become established. Unless, that is, you have a favorite space artist who influenced you in taking your own space tourism flight. In which case let me know, and maybe we can include the award in the first reprint of this book.

Chesley Bonestell

*"I have left the Earth three times, and found no place else to go.
Please take care of Spaceship Earth."*

Wally Schirra, Mercury, Gemini, Apollo astronaut

"Space Tourism? You're kiddin' me right? Only NASA does space – and you need to be a combat-hardened fighter jock to be an astronaut." Such comments would have been the typical response to a discussion about space tourism just a decade ago. We have now reached the chapter where the focus is the new space tourism industry, and where we shall discover just what it took to change those older attitudes. We have seen the early influential players along the road, and even awarded early "Wright Stuff" Awards to those who did lead the way to space tourism from the very onset of aviation and space flight. Now we sharpen our focus. We shall see how a motley assortment of individuals has taken that early promise of space tourism and turned it into reality. Only very few of them will earn the "Wright Stuff" award, however, for their daring vision and commitment to making this happen. All of the others were important in their way, but would probably agree to applaud the recipients, as the true exemplars of the "Wright Stuff" in the space tourists' category. This is where, at last, spaceflight becomes *personal*.

The time scale for this last section is very much shorter than for the previous chapters. That is because the previous chapters have covered the groundwork that was done throughout the previous century starting at Kitty Hawk and Kaluga, both in 1903. Most of the developments in this Chapter have taken place during the last two decades. Your own imminent spaceflight has only just become possible. Only 500 people went into space during the first fifty years of the space age. With the onset of space tourism, the numbers will eventually reach thousands *per year*. But if you fly soon, you may still be able to be among the first 1,000 people in the whole of history to have left the Earth to reach space. And you could even call yourself an astronaut since you will have exceeded the internationally accepted definition of space, which is 100km (62 miles), unless someone changes the definition so that it applies only to those in charge of the spacecraft.

There was an advance guard, which began to present conference papers during the early '90s, and even earlier, which proclaimed the potential economic benefits of space tourism. The account in: "Space Tourism – Do You Want to Go?"(Ref 2) provides details of the very early isolated contributions to the space tourism concept, from as far back as the '70s. People like Pat Collins and Tom Rogers did sterling work twenty years ago, crying out in the wilderness at space congresses (see Ref 1). Tom Rogers (Fig 8-1) was regarded, in author Paula Berinstein's phrase, as the granddaddy of space tourism.

Fig 8-1 Tom Rogers - elder statesman of space tourism.

(Credit: Paula Berinstein)

Name Thomas Francis Rogers

(Credit: Philosophy of Science Portal)

Summary Description
Commercial Space Policy and Space Tourism Booster

Date of Birth
1924, Providence, Rhode Island, USA

Date of Death
February 17 2009, Columbia, Maryland, USA

Nationality US

Achievements
Former Research Administrator at MIT.
Former Deputy Director of Defense Research, Pentagon. Space Station director, Congressional Office of Technology Assessment.

Specific help for Your Ticket to Space
The late, and greatly missed, Tom Rogers was the Founding President of the Space Transportation Association from which bully pulpit he fought and argued and cajoled and shouted about commercial space, and space tourism, while giving testimony on public space travel on Capitol Hill. As Chairman of the family Sophron Foundation, Rogers provided important seed money for many commercial aerospace ventures, including the X-Prize in 1996. He wrote and presented countless white papers and was passionate, creative and energetic. As early as 1977 he asked the then NASA Administrator to begin a program to fly passengers on the Space Shuttle. Tom Rogers was thinking about space tourism long ago; he made sure that everyone on Capitol Hill knew exactly what he was thinking, too.

Tom was a physicist who had worked in the Lyndon Johnson administration and as an administrator at MIT and the Pentagon. This background gave him a healthy disdain for governmental decision processes. He founded the Space Transportation Association and was a frequent speaker on Capitol Hill. He had a direct style that was all his own – once giving the answer: "It's none of your business" to a congressional committee member who wanted to know what a tourist would do in space. Of course, he was making the point that what was being proposed was an essentially American notion – someone pays money for a service, and the service happens to involve going into space. The space tourists could sleep all the time if they so chose; it was their money, their choice. The government only needed to make sure that nobody else would get hurt. Rogers died, aged 85, on Feb 17 2009, one day after Konrad Dannenberg – and the world of space tourism lost two of its pioneers in the same week.

Patrick Collins (Fig 8-2) is a research economist at the University of Tokyo who pointed out in conference papers the problem of insufficient regular satellite payloads in order to achieve economies of scale for the launch vehicle business. In the mid-nineties, he inspired this author to later direct some robust, statistically valid, market research studies of a representative sample of millionaires (known as the Futron/Zogby Studies, appearing in 2002). These studies confirmed the existence of a large potential passenger list of space tourists, even at quite high price points. If human passengers were considered as "payloads", suddenly the solution to the economies of scale problem became apparent. Patrick, like Tom, favors strong language (e.g. "They should be hanged!") when he reflects on the slow response of governments to his eminently reasonable assertions.

Fig 8-2 Patrick Collins - economist and space tourism pioneer.

(Credit: Japantimes.co.jp)

Buzz Aldrin (shown in a 2008 picture in Fig 8-3) has also been a strong and tireless (some might say relentless) advocate of space tourism at least from the mid '90s. He has plenty of awards resulting from his Apollo 11 Moon landing back in 1969, and we cannot present an award to our own Foreword Writer - so we will not present him with one of our precious "Wright Stuff" awards, although he has in truth earned one.

Fig 8-3 Dr Buzz Aldrin, orbital rendezvous expert, space tourism advocate.

(Credit: Amy Sussman)

Name Dr Patrick Q Collins

(Credit: Author. Collins to right in image taken in 2008)

Summary Description
Economist and early space tourism advocate

Date of Birth
1952, Sussex, UK

Date of Death
n/a

Nationality UK

Achievements
Wrote and presented numerous papers extolling the virtues of space tourism. Conducted first market research into space tourism(in Japan) in 1993. Co-founder of Spacefuture.com web-site. Economics Professor at Japan's Azabu university. Astronaut candidate in 1989 for Project Juno (UK). Co-wrote first space tourism book (with David Ashford) in 1990 "Your Spaceflight Manual – How You Could be a Tourist in Space Within 20 years".

Specific help for Your Ticket to Space
Without the early efforts of Patrick Collins and Tom Rogers there would not yet be any such concept as space tourism. Patrick – as early as 1990 - presented a sound economic argument for commercial space tourism to a more academic audience. His argument was based on the need for re-usable space vehicles and the need to derive economies of scale in their use. Patrick, like the author, hailed from the UK, so realized early on that any efforts to support piloted space vehicles would be unlikely to succeed at home. While the author moved to the US, Patrick went to Japan, where, as a result, there is now a strong support for the idea of space tourism.

Rogers, Collins and Aldrin were talking about space tourism in conferences when the usual reaction to the very notion of public space travel was laughter. Perhaps we should give Buzz a special award for persistence and valiant services in presenting the case for space tourism for decades, in the face of misinformed journalists, throughout the media. He even made a rap music record, with the music artist Snoop Dogg, in 2009 to spread the knowledge of space exploration and the "rocket experience" to a wider and ever-younger audience. He had been an early supporter of the visionary space architect John Spencer's Space Tourism Society, from its founding in 1997 (Fig 8-4 and Ref 2). John Spencer is a generous spirit and he has introduced many of the players in the space tourism business to each other. His view of the future is strongly influenced by his many Hollywood connections. As we know everything is possible through the magic of movies, and John therefore thinks it's only a matter of time until it will be possible in real life. The role of his Space Tourism Society, he explains, is therefore to describe the end point, and not worry over much on how to get there. Others can do that. We will get there eventually, and meanwhile we can enjoy the experience via space theme parks, and other kinds of space entertainment, just so long as people are reminded of the ultimate goal.

Fig 8-4 John Spencer - space architect and founder of the Space Tourism Society.

Buzz was everywhere supporting the cause of space tourism, and he effectively becomes a strong link, reaching backward to the successes of government spaceflight of the 1960's and forward to the future of private space tourism in the 21st Century. Aldrin was a Commissioner on the President's Commission on the Future of the US Aerospace Industry in 2002 (Fig 8-5), and in that capacity made early and effective use of the findings of the Futron/Zogby Study.

Fig 8-5 Buzz Aldrin (4th from right) sits on the President's Commission on the Future of the US Aerospace Industry, a.k.a. the Aerospace Commission, 2002.

The author was able to work with Buzz to make Chairman Bob Walker and the other Commissioners aware of the business potential of space tourism. Consequently they agreed to introduce language into the Commission's findings and recommendations (Fig 8-6) about the potential market and business opportunity for public space travel. This was a first in a blue-ribbon report to the President. Recommendation #3 stated:

"The Commission recommends that the United States create a space imperative. The DoD, NASA, and industry must partner in innovative space technologies…[which will] enhance our national security, provide major spin-offs to our economy…and *open up new opportunities for public space travel* and commercial space endeavors in the 21st Century." (Emphasis by author).

By inserting this key language, Buzz permanently removed the "giggle factor" from future discussions about space tourism. It is hard to realize now that space tourism has become a fact of life, that for decades no one, particularly those in government or in management of space companies, would seriously consider the idea. Buzz changed that. He risked being ridiculed, because he believed it was important. He continues to be a champion of lower cost access to space, and to that end has created his Sharespace Foundation to find ways for average folks to get into space.

Fig 8-6 Final Report of the Aerospace Commission, a.k.a. the Walker Report, November 2002.
(Credit: Author/Aerospace Commission)

The results of the survey of space tourism demand amongst millionaires helped Buzz Aldrin make his case to others on his Commission who had not previously given the idea of space tourism any credence. One of the reasons that the survey findings were so compelling was that the hard grind of the market analysis work was delegated to two Futron staff market analysts (Suzy Beard-Johnson and Janice Starzyk) with no preconceived notion of, or belief in, the existence of a viable space tourism market. In fact Janice could not even imagine why anyone would want to do such a thing! Another reason was that an ex Space Shuttle Commander, the safety-conscious Brian O'Connor (Fig 8-7), had taken part in guiding those of us who prepared the questionnaires. He made sure that the millionaire respondents were aware of the realities of space

travel, both good and bad, before they were asked to give their responses. The same questions were asked in several different ways, and reasonability checks were conducted to compare the responses to previous expenditures that the respondents had made for leisure activities. Also, to help make its case, the study pointed out just how many rich persons there were - around 8 million millionaires and 400 billionaires at the time of the study. In the book of Harvard business case studies published in 2008, Ref 5, we can read of the study devoted to space tourism that "Like the shot heard round the world, the Futron study provided the economic incentive for the rocket men to press forward with their dreams." Perhaps the language is a little bit on the extreme side, but this author is not complaining, since the industry has certainly built its business cases upon the revenue expectations developed in those original study forecasts published in 2002.

Fig 8-7 Shuttle astronaut Brian O'Connor added realism to the Futron/Zogby Survey. He is currently Chief of Safety and Mission Assurance at NASA.
(Credit: NASA)

Space tourism began with the most difficult, and therefore most expensive, variant, i.e. orbital space flight. Prices vary, but are generally in the range of $20 million to $35million. The activity began in December 1990 with the Russians, (then still in the Soviet Union), responding to a request

Name Toyohiro Akiyama

(Credit: JAXA/Roskosmos/Spacefacts.de)

Summary Description
Orbital space tourist, 2 December, 1990

Date of Birth
22 July 1942, Setagawa, Japan

Date of Death
n/a

Nationality Japanese

Achievements
First private space traveler, and journalist (reporter for Tokyo Broadcasting System – TBS).

Specific help for Your Ticket to Space
For someone who is decidedly a "first" at something as impressive as going into space as a private citizen, and who moreover is a journalist, Akiyama has remained surprisingly invisible to the world outside of Japan since his 1990 flight. His flight was all the more remarkable when one considers that his newspaper negotiated with not the Russian government, but the Soviet Union, to make his trip to the Mir possible. Nevertheless, it was his flight as the first non-governmental space traveler that provided the first data point for the new industry of space tourism. The experience must have been very disorienting for Akiyama, culturally, physically and politically. Maybe that is why he has remained a rather un-heroic figure for the public leader of a transforming movement as important as space tourism.

from Japan, and offering the first civilian flight for a fee to the Japanese journalist Toyohiro Akiyama (Fig 8-8), on a Soyuz (Fig 8-9) visiting the Mir Space Station. His return capsule (Fig 8-10) is on display at the Washington DC National Air and Space Museum, complete with the cosmonauts' autographs in chalk on the outside written after landing, which is the customary way for cosmonauts to celebrate their safe descent. One of them, of course, is in Japanese Kanji script characters. The museum, however, currently has the capsule tucked away, where it is difficult to see, and most people who do see it probably do not realize that this blackened and burned object was the first space tourism vehicle.

Fig 8-8 Toyohiro Akiyama, first private space traveler, 2nd December, 1990.

(Credit: Spacefacts.de)

Fig 8-9 Soyuz spacecraft of the kind that carries space tourists into orbit.

(Credit: Energia)

Fig 8-10 The first ever space tourism vehicle. Return capsule from the Soyuz spacecraft used by the first private space traveler, Toyohiro Akiyama, 1990. Note Akiyama's chalk signature in Japanese characters on the left side of the capsule.

(Credit: National Air and Space Museum)

This example was followed less than a year later by the British chemist Helen Sharman who had a similar adventure (Figs 8-11 and 8-12). There then followed a decade while the Soviet Union collapsed, the Russians lost their launch site to a neighboring sovereign state, Kazhakstan, and at the end of the decade, in Virginia, USA, a new entrepreneurial company Space Adventures Corporation was formed.

Name Dr Helen Patricia Sharman

(Credit: Stardome-Planetarium.com)

Summary Description
Orbital space tourist, 18 May 1991

Date of Birth
30 May 1963, Sheffield, UK

Date of Death
n/a

Nationality UK

Achievements
First British astronaut, selected from 13,000 applicants for Project Juno. Flew Soyuz TM-12.

Specific help for Your Ticket to Space
Helen Sharman responded to a full-page advertisement in a UK newspaper seeking astronauts. She was successful and flew to Mir in 1991, a year after Akiyama had led the way. The funding for the trip came from a consortium of British companies, so Sharman, like Akiyama, is not a true space tourist in the sense of paying for the flight from one's own pocketbook, or purse. Nevertheless, she did fly in Soyuz as a non-governmental astronaut a full decade before Tito, who is regarded at least in the US as the first space tourist. She pioneered the path for those to follow, including other women such as Anousheh Ansari 15 years later. This was the last such trip during the Soviet era. Soon after her return, ownership of space assets in Russia became complicated, and so the next private flights had to wait a decade. It has always seemed to the author to be an amusing coincidence that Sharman's company, for which she worked as a chemist before going into space, was called Mars.

Fig 8-11 Helen Sharman, first female private space traveler 18th May 1991.

(Credit: Stardome-Planetarium.com)

Fig 8-12 Helen Sharman with Alexei Leonov in Glasgow in 2008. He had directed her training in 1991 during the Soviet era.

(Credit: British Interplanetary Society/www.BIS-spaceflight.com)

Sharman, a chemist at a chocolate confectionery firm, was selected competitively. Funds were raised from sponsorships. The training period was 18 months long. Fig 8-12 shows Sharman later with Alexei Leonov, who was head of cosmonaut training in the Soviet Union when she had undertaken her flight. Helen Sharman flew only 8 years after Sally Ride, who was the first American female astronaut, and she did it with very little support infrastructure. In fact, she was largely on her own pursuing her dream of space in the Soviet Union. She is a gracious and courageous lady, and in December 1992, HM Queen Elizabeth II awarded Sharman the OBE, for being the first Briton to venture into space.

Space Adventures was founded by Eric Anderson (Fig 8-13) and Peter Diamandis (Fig 8-14), of whom we have already heard. Anderson, in his distinctively determined way, went ahead to create the new era of true space tourism where the space tourist actually pays for his/her own ticket (see Ref 4).

Fig 8-13 Space Adventures CEO Eric Anderson, with Buzz Aldrin, at Explorers' Club annual dinner in 2007.

(Credit: Space Adventures)

Akiyama's ticket had been paid for by his newspaper; Sharman's ticket costs were paid by a fund-raising venture in the UK called Project Juno. At the time, the author was a director of a UK body called The Space Society, which gave support to the venture which had been created by two of our members. Project Juno began with a full-page advertisement in the London "Times" newspaper on June 30th, 1989, with the captivating headline: "Astronaut Wanted. No Experience Necessary". From the 13,000 hopefuls who responded,

You already know that tickets for orbital space tourism are expensive. Under Anderson's guidance and entrepreneurial risk-taking a succession of space tourists signed up, paid their fees, were trained, and subsequently went to space aboard Soyuz rockets and docked with the International Space Station. They were all pioneers and should be celebrated as such. Let's give them a round of applause as we think about the risks they took, the money it cost, the six months of exhausting training, and the red tape they had to circumvent. Not to mention having to learn a whole new language, namely Russian (in order that they could perform as a member of a Russian speaking crew particularly in emergencies).

Fig 8-14 Peter Diamandis, Space Tourism vision-ary and entrepreneur.

(Credit: Strategic News Service)

Dennis Tito was the first true space tourist, in April 2001 (Fig 8-15, Fig 8-16). He had to do a great deal of his own international negotiating, and was the first to try out the arrangements that the post-Soviet era Russians were making to allow space tourists to be trained and to fly. He was 61 when he made his flight. Before becoming a financier, Tito had designed spacecraft trajectories at the Jet Propulsion Laboratory. He had to overcome active efforts by NASA to prevent him from flying – but his determination won through. Why did NASA object? They stated at the time that it was about safety, and distractions to the regular astronaut crew. The ISS was, after all, designed as a working laboratory and not as a hotel. But we might suspect that there was also an element of protectionism by the old guard who were concerned about the status of the astronaut elite if just anyone would be allowed to fly into space.

Name Eric C. Anderson

(Credit: Jeff Foust/The Space Review)

Summary Description
Space tourism entrepreneur

Date of Birth
18 April 1975, Colorado, USA

Date of Death
n/a

Nationality US

Achievements
As President and CEO of Space Adventures, Inc., Anderson demonstrated that there was a market for orbital space tourism, even at prices above $20m. He co-wrote "The Space Tourist's Handbook" in 2005.

Specific help for Your Ticket to Space
Anderson was responsible for organizing flights of the first orbital space tourists. Eric Anderson created the "one-stop-shop" where aspiring orbital space tourists could come to arrange their flights. Since all the original flights had to take place using the Soyuz vehicle, this meant creating an operation that could handle the Russian end of the transaction. Since orbital space tourism started before sub-orbital space tourism, it is therefore a truism to say that Space Adventures was the first space tourism travel agency. These early orbital space tourists had to spend around six months in Russia undergoing medical tests and training, including Russian language training. When he created his company, he had to operate initially in an environment, at least in NASA, that was openly hostile to the very idea of sending space tourists into space, and particularly to a space station destination. His initial successes resulted in changes in the official attitude in the US to space tourism using Russian space hardware.

Name Dennis Anthony Tito

(Credit:Space Adventures)

Summary Description
Orbital space tourist, 28th April, 2001

Date of Birth
August 8 1940, New York, NY, USA

Date of Death
n/a

Nationality US

Achievements
Former scientist at NASA's Jet Propulsion Laboratory,
Founder, Wilshire Associates (a numerical approach to Wall Street investment management). First self-financed space tourist

Specific help for Your Ticket to Space
Certainly within US circles, Tito is regarded as the first space tourist (thus perhaps ignoring the efforts of Akiyama and Sharman a decade earlier, who, although non-governmental astronauts, had been funded by their respective commercial organizations). Tito had a tough time fighting the bureaucracies in both Russia and the US in order to be allowed to pay for his flight via Soyuz to the International Space Station. NASA at the time was opposed to the venture. Fortunately, the Russians had a better grasp of capitalism. Every subsequent space tourist benefited from the work that Tito did to enable the training and medical certification to take place in Russia. He sought and received help from both Mircorp and Space Adventures in achieving his ambition.

After the mission, Tito declared how much he loved the experience, and he was invited to address Congress at a hearing of the US House of Representatives Subcommittee on Space and Aeronautics, to discuss his experiences and the future of a potential space tourism business. At the same hearing, Buzz Aldrin added his own comments: "NASA's refusal to actively encourage is a major hurdle." It had taken Dennis Tito 10 years to bring about his flight – he began the process when Akiyama and Sharman were having their flights, but was confounded by the collapse of the Soviet Union. Because of the significance of his flight, Tito's spacesuit is itself now a prized historical artifact on exhibit at the Smithsonian's Air and Space Museum in Washington, DC.

Fig 8-15 Dennis Tito in the International Space Station (ISS), April, 2001.

(Credit: Space Adventures)

Fig 8-16 An exuberant Dennis Tito on landing, after the first true space tourism flight – bought and paid for from his own pocket.

(Credit: Space Adventures)

Tito was followed just a year later by the very much younger South African entrepreneur of Internet encryption software Mark Shuttleworth (Fig 8-17, Fig 8-18). Shuttleworth's business success with the Internet security firm he operated out of his parents' garage meant that he was wealthy enough to pay for the flight at the age of 28. His flight arrangements went more smoothly than was the case for Tito, and his flight followed 8 months of training, based on a training contract he nevertheless had to negotiate himself with over a dozen different groups in the newly emerged post-Soviet Russian space program. This was a tricky time for business dealings with Russia, since much was in disarray, and the Russians themselves did not even know who was responsible for what.

Fig 8-17 Mark Shuttleworth's happy landing.

(Credit: africaninspace.com)

Fig 8-18 Mark Shuttleworth on a space tourism stamp.

(Credit: Wvegter.hivemind.net)

Name Mark Richard Shuttleworth

(Credit:Africaninspace.com)

Summary Description
Orbital Space Tourist, 25th April, 2002

Date of Birth
18 September, 1973, Welkom, South Africa

Date of Death
n/a

Nationality South African/UK

Achievements
Founded Thawte Consulting, Internet security firm, in1995. First African in space.

Specific help for Your Ticket to Space
Almost exactly a year after Dennis Tito, Shuttleworth became the second true space tourist. He even conducted some space experiments from orbit, which he had selected from South African academics, including embryonic stem cell research, cardiovascular system research and crystallization behavior under zero-g. He now works to improve education skills in Africa. It was not until after his mission that Shuttleworth realized that he had never seen a rocket launch, manned or unmanned, before - and that this is probably some kind of record for an astronaut! All tourists like to keep souvenirs from their travels. Shuttleworth kept his spacesuit and bought his Soyuz re-entry capsule.

Greg Olsen was next, in October 2005 (Fig 8-19, Fig 8-20). In 2007, at a meeting of the Space Investment Summit (which took place at the Ritz Carlton Hotel in Battery Park, New York - very convenient for the Wall Street venture capitalist community who were the intended audience) Olsen said that he was known in his home town Princeton as "That Space Guy." This had the advantage that he sometimes received a free bottle of wine at restaurants. Olsen had made $700m selling his Sensors Unlimited Company. It took him, once he had begun the process, a year and a half to get into space, because he had to resolve some medical problems before being allowed to continue his training. By the time of his flight, NASA had begun to change its attitude about space tourists. In fact an American – a NASA astronaut – Bill McArthur, also occupied another of the extra Soyuz seats *en route* to the ISS. This situation had arisen because of the grounding of the entire US Space Shuttle fleet following the re-entry breakup of the Shuttle "Columbia". The same situation will re-emerge after the de-commissioning of the Space Shuttle fleet before the end of 2011. The US astronaut corps will be depending upon rides via the Russian Soyuz for several years after 2010 before a US alternative vehicle is in place. Spare seats for orbital space tourists will therefore become very scarce for a while.

Fig 8-20 Greg Olsen in ISS, October 2005, (note Gagarin's photo on the cabin wall).

(Credit: Mype.co.za)

Olsen was followed a year later by Anousheh Ansari, (Fig 8-21, Fig 8-22, Fig 8-23), then by the Microsoft software architect and Hungarian native Charles Simonyi (twice) in April 2007 and March 2009, and in between by Richard Garriott (October 2008).

Ansari gave a talk in February 2007 in Arlington, Virginia. She reported that she had been a back-up space tourist, but was switched to prime just 3 weeks before her flight when the prime candidate Daisuke Enomoto (from Japan) was replaced due to medical issues. She said that she "came home to a normal, perfect landing, but with a different perspective on life". By the time of her flight opportunity, the training period for space tourists using Soyuz had become more or less standardized at 6 months. Her mixed US/ Iranian heritage caused some minor issues about use of flags on the mission, but her infectious enthusiasm appealed to many who followed her exploits all over the world by means of her blog. Anousheh Ansari's mission patch carried the words: "Imagine – Be the Change – Inspire".

Fig 8-19 Greg Olsen in his spacesuit.

(Credit: Space Adventures)

Name Dr Gregory Hammond Olsen

(Credit: Fritz Retharge/The Gazette Newspaper)

Summary Description
Orbital Space Tourist, 1st October, 2005

Date of Birth
20 April, 1945, Brooklyn, New York, USA

Date of Death
n/a

Nationality US

Achievements
Physicist, created Sensors Unlimited, making opto-electronic devices (near-infrared cameras).

Specific help for Your Ticket to Space
Greg Olsen is physically tall, and so as a result of his mission, adaptations have been made to accept candidates who exceed previous height restrictions (just so long as they can pay the bill for the flight ;-) He was even able to overcome a previous medical condition involving a problem identified in a lung X-Ray, and continue his training. Dr Olsen admits to having been a juvenile delinquent, and now contributes to charities for underprivileged kids. He became the third private citizen to pay for the space tourism experience, although because of the strenuous and lengthy training ("a combination of college and military training") he objected to the "space tourist" description, preferring "space flight participant". Olsen conducted some experiments for ESA while in orbit, including a motion perception study, a lower back pain study, and a microbe collection experiment.

Name Anousheh Ansari

(Credit: John Spencer)

Summary Description
Orbital Space Tourist, 18th September 2006

Date of Birth
12 September1966, Mashhad, Iran

Date of Death
n/a

Nationality US/Iranian

Achievements
Co-founder and chairman of Prodea Systems, Inc. Co-founder of Telecom Technologies., Provider of funding to X-Prize Foundation , First self-financed female space tourist, First Iranian astronaut

Specific help for Your Ticket to Space
With her brother-in-law Amir, Ansari supported the X-Prize Competition with funding. It thus became the Ansari X-Prize. Anousheh Ansari has therefore supported the creation of a sub-orbital space tourism industry, while taking part as an orbital space tourist herself. Ansari is labeled the world's first private female space explorer, but that title rightly belongs to Helen Sharman. Ansari is, however the first *self-funded* female private space explorer. Ansari was the first private space tourist to actively blog before and during her space flight, thus involving many young people from all over the world in real time . She was the first space tourist back-up to replace a prime flight candidate. Ansari is a powerful advocate of space exploration and space tourism (see photo above!).

Fig 8-21 Anousheh Ansari prepares for her orbital flight.

(Credit: Space Adventures)

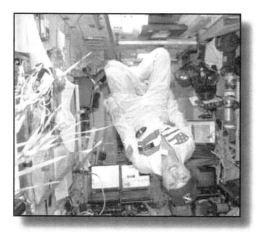

Fig 8-22 Anousheh Ansari in the ISS, wearing both US and Iranian flags, September 2006.

(Credit: Space Adventures)

Fig 8-23 Ansari and her crew after safe landing.

(Credit: Space Adventures)

Charles Simonyi, who made his fortune through helping develop Microsoft Word and Excel, as you might expect managed to negotiate a "bulk buy" deal, so that his second flight was significantly cheaper than his first trip. When asked why he wanted to fly a second time, Simonyi said that he wanted to improve his skills (Fig 8-24, Fig 8-25). As a 13-year-old in his native (then-Communist) Hungary he had made a "Junior Cosmonaut" trip to Moscow. Simonyi had enthusiastically followed the early developments of the Soviet space program and could even relate the names of the various dogs that had flown on missions prior to Gagarin. He left Hungary at age 17 for the United States in 1968. It seems ironic that he had to leave Hungary to make his fortune in the US in order to be able to go back again to the launch site in the former Soviet Union about which he dreamed as a boy. Some people are simply very determined to realize their hopes and dreams. On the way to achieving his space flight opportunities, Simonyi had learned to fly and he is a private pilot.

Richard Garriott, who is an online computer games entrepreneur, and developed the successful Ultima and Tabula Rasa games, was the first "second-generation" astronaut. His father Owen Garriott had flown as a NASA astronaut aboard the Skylab mission in 1973, and on early Space Shuttle flights. Fig 8-26 presents Richard Garriott with Space Adventures' CEO Eric Anderson. Garriott was born in the UK and has joint UK/US citizenship. Much of his computer gaming has a British theme. He grew up in Houston in an environment where all of his neighbors were astronauts, and he even owns a rare full-scale engineering model of Sputnik 1. Richard Garriott is an angel investor in Space Adventures, as well as in other entrepreneurial space tourism companies like XCOR Aerospace, so he clearly believes in doubling or tripling up on his risks where space tourism is concerned.

Fig 8-24 Charles Simonyi (who flew twice as a space tourist) in his Russian Sokol spacesuit.

(Credit: spaceadventures.com)

Fig 8-25 Charles Simonyi after recovery from his first space tourism flight, April 2007.

(Credit: NASA/Bill Ingalls)

Name Dr Charles Simonyi

(Credit: Space Adventures)

Summary Description
Orbital Space Tourist, 7th April 2007 and 26th March 2009

Date of Birth
September 10 1948, Budapest, Hungary

Date of Death
n/a

Nationality US/Hungary

Achievements
Microsoft Word and Excel software entrepreneur
First orbital space tourist to fly twice.

Specific help for Your Ticket to Space
Simonyi proved just how wonderful the process of space tourism is by going up twice within two years. Maybe his decision to duplicate his first space tourism flight stems from his days after he arrived from Hungary and started working at Xerox Corporation. Maybe he was just hungry for more. Even though he was taking advantage of a cannily-negotiated bulk discount, this was nevertheless an extraordinary step to take with prices generally over $20M per trip at the time, and he reported "missing showers and beer" during his stay on the International Space Station. Simonyi provided full web-site interconnectivity during his flights, and he answered over 100 questions from the public about life in space while in orbit. The site had 4 million visitors. He conducted experiments on radiation and biological organisms while at the ISS.

Name Richard Allen Garriott

(Credit: Space Adventures)

Summary Description
Orbital space tourist, 12 October 2008
First 2nd Generation astronaut (father Owen Garriott)

Date of Birth
4 July 1961, Cambridge, UK

Date of Death
n/a

Nationality US/UK

Achievements
Computer games designer, Ultima and Tabula Rasa.
Donor and board member of X-Prize, Zero-g Corp,
XCOR and Space Adventures.

Specific help for Your Ticket to Space
Richard Garriott grew up surrounded by astronauts.
His father Owen Garriott went up in one of the
last Apollo spacecraft to perform his duties on the
Skylab – an early US space station. Richard intended
becoming an astronaut himself when he grew up.
Real life intervened in the shape of a NASA doctor
who told him he would not make it due to an eye
condition. So he found another way. He made
his fortune designing computer games, and was
therefore able to pay for his own trip, arriving at the
space station 35 years after his father's own trip to an
earlier space station. He even had a liver operation so
that he could qualify. Garriott conducted experiments
on crystal growth, and talked with schoolchildren
from orbit. Garriott is a great student of Shakespeare
(and even has his own personal copy of the Globe
Theater at his Austin, Texas home he calls "Britannia
Manor"), and now he *really* knows what it means to
say "All the world's a stage".

Fig 8-26 Eric Anderson of Space Adventures
(foreground), with space tourism client Richard
Garriott and a Russian space suit.

(Credit: Space Adventures)

Fig 8-27 tracks Richard Garriott as he
approaches his Soyuz rocket, accompanied by his
father. It was ironically only 5 years earlier when
the author talked with Owen Garriott, who was a
fellow speaker at a Washington DC Conference
where space tourism was being discussed. Owen
Garriott back then declared that: "…they will
like space tourism, and generally people will tell
their friends – but I don't believe your forecasts
for tourists willing to pay $20m to go orbital,
though!"

Fig 8-27 Owen Garriott, former Skylab astronaut,
accompanies his son Richard to his launch vehicle
for his space tourism trip, October 2008.

(Credit: Energia)

And then Guy Laliberté, the founder of the Cirque du Soleil, (Fig 8-28, Fig 8-29) had his flight, rounding off the orbital space tourism circus just as the sub-orbital space tourism business was about to begin operations. Laliberté has been an accordionist, stilt-walker and fire-breather. He is clearly not averse to risk-taking and he won almost $700,000 at a Poker Championship at the "Bellagio" casino in Las Vegas before he undertook his flight. Once in orbit at the ISS, he managed a global entertainment event with a socially conscious agenda. He has a personal aim to fight poverty in the world by giving everyone access to water. He took up nine red clown noses to space – one for each member of the ISS crew.

Fig 8-28 Guy Laliberté plays professional poker before undertaking his orbital space tourism adventure in September, 2009.

(Credit: Lasvegasvegas.com)

Fig 8-29 Guy Laliberté in the ISS October 2009.

(Credit: NASA)

Name Guy Laliberté

(Credit: Space Adventures)

Summary Description
Orbital space tourist, September 30th 2009

Date of Birth
2 September, 1959, Quebec City, Canada

Date of Death
n/a

Nationality Canadian

Achievements
CEO of Cirque du Soleil

Specific help for Your Ticket to Space
If Laliberte is a clown, he is a very wise one indeed. Guy Laliberté was able to pay $35M for his orbital mission. The Russians were monopoly suppliers of seats to orbital space tourists, and the US official human space policy was undergoing review, with the probability that the US Space Shuttle would stop flying in 2010/2011. Furthermore, it was unlikely that there would be many more openings for "spare" Soyuz seats, since the US government would after 2010 be buying them to launch their own astronauts to the International Space Station for a period of several years. Since the US government has been charged $51m for each of *its* astronaut's Soyuz seats, it seems that Laliberté's years as a professional poker player, and managing high wire acts, were not wasted! Whilst in orbit, Laliberté, who had created his "One Drop" Foundation, orchestrated an "artistic and poetic mission" which brought attention to water issues on Earth.

It is somewhat striking that many of these multi-millionaires, who went into space via Space Adventures in the first decade of the new millennium, were entrepreneurs from the financial, Internet, or IT industries. And of course, it is noticeable that while 5 of them were US citizens, their flights were conducted from the steppes of Kazhakstan at Baikonur, and not from Kennedy Space Center, or any other base in the USA. Eric Anderson, as the entrepreneur whose efforts and commitment created the first successful space tourism company, receives a "Wright Stuff" award for his vision, risk-taking and perseverance in the face of so many obstacles.

Eric Anderson

A new stage of space tourism was about to be born, however, and this time it would not be the Russians at its forefront. The US was about to bring into place a very much lower cost variant on public human spaceflight; the sub-orbital space tourism experience, which would open up space tourism to thousands of eager travelers like yourself (Ref 6). This possibility had been explored by the 2002 market studies, and the findings suggested a significant market awaited the technology if prices could be brought to about $100,000. We have seen that Buzz Aldrin had included this information that year in the Report of the Presidential Commission of which he was a member. Sub-orbital space flights, such as yours, would in effect be a repetition of the experience of America's first astronaut Alan Shepard. The main difference would be that they would involve taking off and landing at the same land base, whereas Commander Shepard had the entire Atlantic Fleet of the US Navy awaiting his return from space. New laws and regulations would be needed.

The military space centers were unsuitable for this new kind of activity, so a new series of civilian "spaceports" were brought into operation.

Under the George W Bush administration, a part of the Federal Aviation Authority (the Office of Commercial Space Transportation, or FAA-AST) was given responsibility for providing the regulatory environment for the new industry.

The first of these new kinds of spaceport was Mojave, California, home to Burt Rutan's Scaled Composites venture. Its General Manager is Stu Witt, who has ensured that his facility continues to be ideal for test flying. Stu has a steely resolve and a no-nonsense manner.

At the FAA-AST, Patti Grace-Smith was since 1998 the very effective, Associate Administrator for Commercial Space Flight (see Fig 8-30 for an image with Patti taking part in X-Prize Cup celebrations). Patti is now a consultant who brings a very thoughtful, yet practical and optimistic, perspective to her work. She strives for a balanced view that still reflects the concerns of her former role, namely the safety of the uninvolved public, while encouraging the development of the new space tourism industry. Patti serves on the new (2010) Commercial Space Committee to the NASA Advisory Council. Her successor at the FAA-AST, under the Obama Administration, is Dr George Nield, whose responsibilities after 2010 will include the regulation of commercial flights to the ISS carrying US government astronauts, orbital space tourists, or a combination.

Fig 8-30 Patti Grace-Smith, center, with Peter Diamandis, Buzz Aldrin, Rick Holman (New Mexico Economic Developments) and a Wirefly executive at the Wirefly X-Prize Cup in Las Cruces, New Mexico, 2006.

(Credit: Jeff Foust/The Space Review)

In order to make these activities possible, throughout 2003 and 2004 work had to be done on the Hill to bring about the passage of the Commercial Space Launch Amendments Act (CSLAA) of 2004, which enabled the space tourism regime in the US to be established. It extended the licensing requirements of the original Commercial Space Launch Act of 1984. The drafters were well aware of the risks in this industry. But they also knew that if too much regulation had been imposed a century earlier, the Wright Brothers would not have succeeded in setting in motion the aviation industry we have today. They recognized what we already know, that space tourism was of the essence of the American spirit. Perhaps the single individual most responsible for the passing of the bill was Jim Muncy (Fig 8-31), a former Hill staffer and regulatory consultant, and a founder of the Space Frontier Foundation, who worked tirelessly for its passing. His is a familiar figure at commercial space conferences, often, since he does not need a microphone to be heard, providing commentary from the "peanut gallery" at the back of the room. Even he, however, was surprised to learn that the bill had passed, late at night. He had been exhausted and went to bed thinking it was a lost cause, and awoke to find that it was now the law of the land.

The new law indemnified the US Government from any personal injury lawsuits, established the FAA as the regulatory agency, insisted on space tourists being fully informed of the risks, and in effect set up the framework for a space tourism regime and its associated insurance requirements.

Fig 8-31 Jim Muncy – smoothed passing of critical space tourism law at the US Capitol. (Photo Nov 2005).

(Credit: Students for Exploration and Development of Space)

Another "mover and shaker" on the Hill had been Charles Miller of the Space Frontier Foundation, a commercial space activist and entrepreneur who organized an annual "March Storm" event, when volunteers would turn up in congressperson's offices to present the case for commercial space (Fig 8-32). Under the Obama Administration, a number of key places in NASA leadership would go to the former activist commercial entrepreneurial outsiders, and they would bring about a major change in direction for the country and at the Agency. Charles Miller is one such new entrant now working at ensuring that NASA technology developments in future are beneficial to the whole space manufacturing and operating community, including the entrepreneurial startups. He draws his parallels from the early days of NASA's predecessor organization NACA, which developed such universal items as engine cowling designs and aerofoil data that the whole industry could use. The author worked with Miller when he was establishing an entrepreneurial space cargo delivery company, CSI, and trying to win a NASA contract for delivery of payloads to the ISS. It is a significant change to the NASA management when a former space entrepreneur is now inside the doors, helping to set direction. One still expects him somehow to bring out a protest banner, or maybe unfurl a pirate flag, during his NASA presentations!

Fig 8-32 Charles Miller at the podium. He organized grass roots support for space commerce laws and now works inside NASA to bring about further change.

(Credit: Jeff Foust/ The Space Review)

A major force for commercial space and space tourism has been the appointment of Lori Garver (Fig 8-33), as the new deputy to the new NASA Administrator Charles Bolden. Lori had previously been so committed to the concept of space tourism that she was for a while herself a candidate for a Soyuz flight, and she underwent much of the training before funding dried up and she could not continue. She is a former Executive

Director of the National Space Society. There has certainly been a "sea-change" at NASA when we recall the battles that Dennis Tito had to fight with the Agency in order to even be allowed to undertake his 2001 flight.

Fig 8-33 Lori Garver with the author and Peter Diamandis at a meeting of the Washington Space Business Round Table in January 2003. Garver was appointed NASA's Deputy Administrator in 2009.

(Credit: Jeff Foust/The Space Review)

Administrator Charlie Bolden, a former Space Shuttle astronaut and Marine, was a formal reviewer of the National Research Council's seminal 2009 report: "America's Future in Space - Aligning the Civil Space Program with National Needs". He completed his reviewer duties just before being sworn into his new position, where he would be able to implement many of the report's recommendations. Such as, requiring a big increase in advanced space technology R&D, expanding international partnerships, and seeking to provide economic and societal benefits that contribute solutions to the nation's most pressing problems.

With the CSLAA Act in place (and subsequent supportive leadership within NASA), the FAA-AST would be able to approve the new spaceports, and establish the regulatory regime for the new industry, including such matters as training requirements. Now everything was in place for business entrepreneurs to proceed and make their money from space tourism. They knew the markets were there, and now they had a supportive regulatory regime. The main premise of the CSLAA law of 2004 was that it was no responsibility of the US government to prevent people from doing inherently risky things, provided that the uninvolved general public was protected. Tom Rogers had said it more bluntly a decade earlier. This was an acceptable premise, and future

passengers such as yourself would simply have to sign indemnification statements before their flights. Only one problem, though. The technology for providing the public with sub-orbital space tourism experiences did not yet exist.

This is where the brilliant Peter Diamandis again drives the story. We have seen that Diamandis had already, in 1988, founded the International Space University, which provides graduate training courses on space flight and space business to students from all over the world. He had been a founder of Space Adventures, too. And he was working at establishing another company called Zero-G Corporation, for providing weightless experiences to the public in aircraft flying in parabolic trajectories.

Peter Diamandis

Diamandis's next great idea in 1996 was the X-Prize, of which we have been hearing, subsequently called the Ansari X-Prize when he obtained the bulk of necessary funding from the Ansari family. He had read Lindbergh's autobiographical book, and decided to follow the idea, from the early days of aviation, of awarding a prize for developments in space tourism. He wrote simple rules. The prize of $10m would be awarded to the first private craft that could reach the boundaries of space (defined as 100 km or 62 miles), carrying three persons (or their equivalent weight), and then repeating the feat within two weeks. The winning team would have to privately finance, build and launch a spaceship. And then launch it again within a fortnight. If someone could do this, he reckoned, then this would demonstrate re-usability and open up the new space tourism business. We shall read later how the competition worked out, but at this point we have no hesitation in awarding Diamandis one of our "Wright Stuff" awards for enabling space tourism. He has not finished his pioneering work, but went on to create the X-Prize Cup annual event, the Rocket Racing League and the Google Lunar X-Prize competition (see Ref 5). He founded the Singularity University as recently as 2009, which has aims of achieving some major breakthrough thinking in many sectors, including medicine.

We pointed out how Buzz Aldrin has supported space tourism. Although not to the same degree, several other Apollo astronauts have also embraced the call of public space flight. Future space tourists like yourself should know that they continue in a line of pioneering space endeavors linked to the very beginning of the space program. There are some very strong links to the early years that are in the direct line of space tourism developments.

Look for example at the sign outside a British public house in Fig 8-34. The tavern is located in Hucknall, near Nottingham, close to the Rolls Royce factory which created the strange vehicle shown on the sign, known as the Flying Bedstead.

Fig 8-34 British pub sign celebrates predecessor of vertical landers.

(Credit: www.urbed.coop)

This vehicle was built and flown in 1954, and demonstrated hovering techniques which were subsequently used in the British Harrier aircraft, in the Lunar Landing Research Vehicle (LLRV) in which Apollo mission commanders could practice their upcoming Moon landings, and in the Lunar Lander (or LEM) itself. It would emerge that the approach would also subsequently be adopted for some space tourism vehicle designs. Fig 8-35 records astronaut Pete Conrad, who would subsequently land on the Moon in Apollo 12, flying the LLRV.

Name Dr Peter H. Diamandis

(Credit: Zero G Corp)

Summary Description
Serial space entrepreneur

Date of Birth
20 May 1961, New York, NY, USA

Date of Death
n/a

Nationality US/Greek

Achievements
Founder International Space University (ISU).
Founder Students for Exploration and Development of Space (SEDS).
CEO Zero Gravity Corporation.
Founder X-Prize.
Co-Founder Space Adventures.
Architect of Rocket Racing League.
Creator of Annual X-Prize Cup event.
Creator Google Lunar X-Prize.
Founder Singularity University.

Specific help for Your Ticket to Space
All of the above! Diamandis is perhaps the one single individual who could claim to have made space tourism happen. Sure, he needed others, like Burt Rutan. But it was his vision and relentless efforts to push the concept, and translate words and ideas into practical consequences, that gave us the Ansari X-Prize events of 2004. Most ordinary folks would have been well satisfied with doing any one of the items listed above in their career. Diamandis never rests, however, and keeps coming up with new ways to move society forward. He declares that his motto is "The best way to predict the future is to create it yourself!"

Fig 8-35 Pete Conrad practices for his Apollo 12 Moon landing in the LLRV.

(Credit: NASA)

Fig 8-36 presents Conrad's crew of Dick Gordon and Alan Bean. We have already met Bean in the previous chapter on artists, and noted his efforts since Apollo 12 at communicating the spaceflight experience to new generations via his artwork and books.

Fig 8-36 Crew of Apollo 12 – Pete Conrad, Dick Gordon, Al Bean, space tourism supporters all.

(Credit: NASA)

Dick Gordon, the Apollo 12 Command Module Pilot, has also supported the space tourism business, as can be seen from the comments he added for the author in Fig 8-37, which is a photo he took in Lunar orbit. It shows the Lunar Lander carrying his colleagues Conrad and Bean to the moon's surface, and it clearly has many design features in common with the Flying Bedstead. This kind of control technology continues to interest space tourism advocates, as we shall see later.

Fig 8-37 Dick Gordon photographs his colleagues descending to the lunar surface in their Lunar Lander during Apollo 12 mission in November 1969. Lunar Lander technology is now being adapted for some space tourism vehicles.

(Credit: Author)

Pete Conrad was a regular speaker at early space tourism conference events. He had a great sense of fun – even mischief - and boundless energy. Someone said he lost at least 10 minutes during his Apollo 12 Moonwalk just through laughing. On landing on the Moon, the five feet six Commander declared: "That might have been one small step for Neil, but it was a big one for me!" The author shared a lunch table with him at a conference hotel near Los Angeles airport in November 1997, and told Pete about having just the previous day become the first life member of the Space Tourism Society. He was delighted to hear news of the new organization, and said: "Whatever it takes. We gotta get things changed somehow. I wanna fly cargoes and passengers on real commercial launches. Anywhere on the planet in 45 minutes or less!"

He died just over a year later crashing his Harley, at age 69. After leaving NASA, Pete Conrad had worked on the DC-X vehicle (Fig 8-38), which used similar hovering technologies to the Lunar Lander, before it ran into funding problems. Pete had acted as the ground-based pilot who flew the DC-X by remote control. Pete had then created his own company, Universal Space Lines, with Dr. Bill Gaubatz, to pursue his point-to-point dream.

Silicon Valley software entrepreneur. He intends to create a successful space tourism business, and regards these precursor vehicles which he tests at the X-Prize Cup events, as his prototype and engineering models. He is competing in a competition known as the Northrop Grumman Lunar Lander X-Prize Challenge, and it rewards the contestants for precision in flying, in vertical takeoff, translation and landing. It is another of Diamandis's ideas, funded under NASA's Centennial Challenges program. Armadillo won $850,000 in prize money spread over 2008 and 2009 for its achievements. Of all the existing major aerospace giants, Northrop Grumman has so far shown the most interest in supporting the development of a space tourism industry. In addition to providing funding for this competition, Northrop Grumman has acquired Burt Rutan's Scaled Composites company, while continuing to allow Burt to operate it as before. Since in 2010, the Obama administration and NASA have introduced a new commercial mandate for human spaceflight, we may see increasing involvement from Boeing and Lockheed Martin, too, in the months and years to come.

Fig 8-38 Pete Conrad was the remote pilot for the McDonnell Douglas DCX (Delta Clipper Experimental) vehicle, which demonstrated RLV technologies for space tourism in 1993.

(Credit: US Army WSMR)

Pete Conrad's dream does go on, however. Fig 8-39 demonstrates the space tourism test vehicle of the entrepreneurial company Armadillo Aerospace operating at an X-Prize Cup event in 2008, and the origins of its motor and control technologies go clearly back to the early years of the Flying Bedstead and the Lunar Lander. Armadillo is owned and managed by John Carmack (Fig 8-40), who is a youthful millionaire

Fig 8-39 Armadillo's "Pixel" vehicle demonstrates hover technologies.

(Credit: Jeff Foust/The Space Review)

Name John D. Carmack

(Credit: Armadillo Aerospace)

Summary Description
Space tourism entrepreneur

Date of Birth
August 20 1970, Kansas, USA

Date of Death
n/a

Nationality US

Achievements
Software entrepreneur, 3D graphics (Doom, Quake).
Created Armadillo Aerospace.
Won first Lunar Lander Challenge with his prototype vehicle.

Specific help for Your Ticket to Space
Like all space tourism entrepreneurs, Carmack is keen to offer his customers a space adventure experience. The current design being developed for sub-orbital space tourism allows for almost unrestricted 360 degrees vision from a vehicle going on a straight up-and-down journey to space and back. Armadillo has been testing out parts of their technology each year at the X-Prize Cup events in New Mexico, where a Lunar Lander Challenge has been taking place. This is because the kind of space tourism architecture which Armadillo is developing is equally suited to landing on the Moon as to performing sub-orbital space tourism flights. Perhaps Armadillo Aerospace will have the least amount of development effort required when space tourism moves onwards to a Lunar destination.

Fig 8-40 John Carmack, founder and CEO of Armadillo Aerospace, with an assortment of rocket spares.

(Credit: Armadillo Aerospace)

Fig 8-41 Masten Space "Xoie" vehicle at the 2009 Northrop Grumman Lunar Lander X-Prize Challenge, where it won $1.1M in prizes.

(Credit: Jeff Foust/The Space Review)

Three other startup companies have demonstrated similar technologies: Masten Space Systems, Unreasonable Rocket, and Bon Nova. Masten Space has, like Armadillo, won prize money for its demonstration of vertical take off, translation and precision landing at a Northrop Grumman Lunar Lander X-Prize Challenge competition, picking up $1.1M in the 2009 competition with its "Xoie" vehicle (Fig 8-41). David Masten, the CEO, is an IT engineer who previously managed major projects for Cisco

Systems and AT&T. It transpires that software development and quality control is a major element required for the success of these vehicles.

Another original Apollo Moonwalker who has worked hard at bringing space tourism into existence is Harrison Schmitt, who landed on the Moon in December 1972 in the last Apollo lunar mission, Apollo 17 (see Fig 8-42). His fellow astronauts called him Dr Rock, because he was a qualified geologist, and on the Moon he had discovered, amongst the all-pervading grey, some orange soil: "It looks just like an oxidized desert soil, that's exactly right!" he reported excitedly from his Taurus Littrow location on the Moon.

Fig 8-42 Harrison (Jack) Schmitt on the lunar surface during Apollo 17, December, 1972. Schmitt has been a supporter of space tourism for over a decade.

(Credit: NASA)

In the preceding photo (Fig 8-43), we see Schmitt chairing a meeting of the NASA Advisory Council in 2007, as Neil Armstrong makes a point about the need for an upgraded air traffic control system to, amongst other things, take space tourism into account. There would be an increasing number of space tourism vehicles zipping up through the atmosphere, many of them descending as gliders, and they had to be integrated into an already-stressed Air Traffic Control system. Armstrong was concerned that not enough effort was being focused on the general issue of upgrading the ATC system to handle the future growth. The author talked with Armstrong during a break in the proceedings, and provided him with the latest forecasts of space tourism numbers for integration with aircraft fleet projections. Armstrong said he was concerned that the ATC upgrade project was a "staggering challenge", and there did not seem to be clear leadership at the time to manage the process going forward. He continues to be concerned about his special field of aeronautics, and the ability of the ATC processes to continue to keep it safe, and he told the author: "It's a humongous problem. It's going to take a long time." From someone who had said, "That's one small step for a man, a giant leap for mankind", these remarks are a little sobering. Fig 8-44 records the author's previous fleeting meeting (about 2 seconds!) with Armstrong in London during his round the world tour following the first lunar landing. Armstrong is an engineer at heart, and is more comfortable performing analysis than, having reached a conclusion, guiding the resulting decisions through the political process.

Fig 8-44 Neil Armstrong outside the American Embassy in London on October 14th, 1969. During the world tour following the historic Apollo 11 mission, the author was part of a crowd who waited 3 hours to greet the crew.

(Credit: Author)

Fig 8-43 Harrison Schmitt (left) with Neil Armstrong at a NASA Advisory Council meeting in 2007.

(Credit: NASA)

Schmitt, by contrast, is a master politician, an effective and witty committee chairman, and former Senator. As a politician, he knows the importance of timing and budget, and choosing the right battles. Schmitt has been a regular attendee and speaker at space conferences, writing papers on use of lunar resources and on space tourism, such as his "Business Context of Space Tourism" contribution at the 2003 Albuquerque Space Technology and Applications Conference. In various discussions with the author, between 1996 and 2008, Schmitt stressed the need to ensure that the insurance community would be available when needed to support the new industry. He also weighed in on the need, or otherwise, of wearing full pressure spacesuits, for descent back to the Earth, especially after long missions.

Other Apollo-era supporters of space tourism include the jocular and easy-going Charlie Duke of Apollo 16, (Fig 8-45). He had described his experience from the Moon with the emphatic word: "Spectacular!" Duke was present in Washington DC on 18th December 2003 at the Wright Brothers' Centenary celebration conference in the Ronald Reagan Building. After the presentations, Duke, who seems like a man very much at peace with himself, told the author of his enthusiasm for the idea of space tourism, and that he was convinced that demand will grow quickly as soon as the first tourists get started: "Oh yes. For sure. They will love it and tell their friends – Hey you gotta do this! It's just a wonderful experience." He did caution, however, about motion sickness, which had been a problem even with professional astronauts: "For public space travel they will need barf bags!"

John Young, the generally low-key and laconic Commander of Apollo 16 (Fig 8-46), and Ed Mitchell of Apollo 14 (Fig 8-47) both told the author of their support of the idea of space tourism. From the Moon, Young had provided the possibly least valuable piece of data to Houston when he announced: "Even the craters have craters!" Young said, in characteristic fashion in April 2003, during a public meeting at the National Air and Space Museum in Washington, DC in answer to the space tourism question: "Yes, I think it's a great idea. You know, for years I've wanted to put a module in the back of the Shuttle,

where you could haul maybe 25 or 30 people up there like in the back of a bus. Then you open the bay doors, and you have windows, and you just fill the thing up with barf bags!" It seems that the Apollo 16 lander crew of Duke and Young share the same focus. One assumes that their mission, or its training, must have provided reasons for this common and clearly articulated view!

Fig 8-45 Space tourism supporter who walked on the Moon – Charlie Duke.

(Credit: Author collection/NASA)

Fig 8-46 John Young, jumping in the Moon's 1/6th gravity, during the Apollo 16 mission in April 1972, is an advocate of space tourism- so long as the space sickness issue can be handled.

(Credit: NASA)

Mitchell's focus was somewhat different. On the Moon, he and Alan Shepard had lost their way. It turns out it is very difficult to navigate on the Moon, because there are no trees or buildings to give scale, and no atmosphere that would be needed to make the distant objects appear relatively indistinct. And of course, there was not even a GPS system on the Earth back then, let alone on the Moon. "Doggone it, you can sure be deceived by slopes here!" he had said in frustration at one point. After he returned to Earth, however, he picked up a clear and distinct new direction. He is a very serious and thoughtful person, much troubled by aspects of mankind's behavior on the planet. He wants as many people as possible to be able to see the Earth from space, because he believes that there is a very positive effect that results from spaceflight, which he calls the "Overview Effect", which leads to a greater appreciation of the planet by the traveler. Mitchell's own experience as a space traveler convinced him of the need for the occupants of Planet Earth to think more globally and treat Earth's resources in a more sustainable way. He wrote in 2006 "We need to approach colonizing beyond Earth as a cooperative global endeavor. I will not cease supporting efforts that lead in the direction of our continued progress." Regarding space tourism specifically, Mitchell continued "Hopefully, we can accelerate the "Overview Effect" by getting a few notables out to look back...and recognize the interconnected web of life". Ed Mitchell grew up in Roswell, New Mexico, and used to walk past Goddard's place on his way to school. He wrote to the author in 2009: "Space tourism will bring great benefits". The author agrees.

Now let's get back to the X-Prize. Once Diamandis had announced the prize, he set about raising the money to be able to write the check for the subsequent winner. This was not an easy thing to do, because, as is usual with winners of "Wright Stuff" awards, they are venturing into risky territory, where no-one has been before. Corporate funders did not want to put their name against something totally unknown and unproven, and which might even be the cause of fatalities. That would not be good for business. Diamandis felt that the old American ideals that rewarded people who pushed back barriers had perhaps

been lost. And then the Ansari family (Anousheh and her brother Amir) stepped forward. They were new Americans, and perhaps for that reason still possessed the pioneer spirit. They had originally come from Iran, and had made their money, like so many others in this space tourism story, in Silicon Valley. They gave their backing to Diamandis's idea, and as we have heard, the Prize was re-named the Ansari X-Prize.

Fig 8-47 Apollo 14 Moonwalker Ed Mitchell is a space tourism supporter. He believes the more people who see the Earth from space, then the more they will realize that the planet is a precious entity that needs to be protected.

(Credit: Author collection/NASA)

We should also record the valiant background services of Gregg Maryniak, who was the Executive Director of the X-Prize Foundation during that high-risk period after which the $10M prize had been announced, and during which there was not yet the money with which to pay it. Other enthusiastic supporters at that time included Ken Davidian who worked at the X-Prize organization. He subsequently spread the prize idea to NASA where the Centennial Prizes were introduced, which led to competitions for robotic challenges amongst other endeavors. He is now bringing his talents and enthusiasm to the FAA-AST, managing the program to create its Center of Excellence for Commercial Space Transportation. You cannot listen to Davidian without being caught up in his infectious enthusiasm. Veteran space tourism journalist Leonard David has been a valued chronicler of space tourism developments since the very beginning, and Jeff Foust has continued

Name Elbert (Burt) L. Rutan

(Credit: Apogee Books)

Summary Description
Designer and builder of aircraft and spacecraft

Date of Birth
June 17 1943, Estacada, Oregon, USA

Date of Death
n/a

Nationality US

Achievements
Designer of aircraft homebuilt kits Vari-Eze, etc. Designer and builder of world record breaking craft Voyager and Global Flyer. Designer and builder of SpaceShipOne and mother plane White Knight, and their follow-ons SpaceShipTwo and White Knight Two (for Virgin Galactic).

Specific help for Your Ticket to Space
Everyone in the space tourism business knows that Burt is the man! He demonstrated that with a small team of about 30 people, and about $20m of private investor financing, it was possible to create a space program, something that previously had only been done by governments. SpaceShipOne went up into space twice within a two week period, in 2004, something that had never been done before, even by governments. Burt is a believer in doing things, not just talking about them. If you want to know what he thinks is government's record in this context, then he will be very happy to let you know with clarity and forcefulness.

the tradition with his own very solid analysis and photographs throughout various publications and blogs in the 21st century. You will have seen many of Jeff's photos in this book.

Twenty different groups registered to compete for the Ansari X-Prize, and it was an important part of Diamandis's concept to leave the choice of architecture unconstrained. Some of the proposals required that the vehicle would take off from land vertically, others horizontally. Some used water for landing and takeoff. Some used balloons. Another part of Diamandis's vision continues with the fact that some of the losers of the competition (such as Rocketplane and Armadillo) are still developing their concepts, and may yet be able to bring them to a public offering.

However, there was only one winner of the Ansari X-Prize in 2004, and that, as we have seen, was Burt Rutan's SpaceShipOne (see Ref 3). Fig 8-48 establishes the always-cheerful Anousheh Ansari alongside the craft in its hangar at Mojave. The craft had been powered by a hybrid rocket motor developed by the company SpaceDev, now Sierra Nevada Corporation, owned and operated by the late Jim Benson, another stalwart supporter of space tourism. Perhaps the most crucial innovation conceived by Burt Rutan was the concept of the "feathered wing". At the maximum altitude, while in the vacuum of space, the pilot "feathers" the wing, and the craft effectively bends in half. The spacecraft becomes anything but streamlined, and can then return to Earth somewhat like a shuttlecock with quite a bit of buffeting but obviating the need for the complex ablative materials used in the Alan Shepard Mercury craft at the outset of the space program. Ablative materials are deliberately burned away sacrificially when heat is applied, as in re-entry at high velocity from space. But therefore they are not suitable for a reusable spacecraft, since between every flight a new ablative surface would have to be built into the craft. The private financing for Burt Rutan's Ansari X-Prize bid had come from Paul Allen, who made his personal fortune from Microsoft. Fig 8-49 shows Paul Allen with Burt Rutan and Brian Binnie, who had flown the second and winning Ansari X-Prize flight in SpaceShipOne. Also in the image are Anousheh and Amir Ansari and Sir Richard Branson.

Fig 8-48 X-Prize funder and subsequent cosmonaut Anousheh Ansari with SpaceShipOne in the hangar of Scaled Composites at Mojave spaceport in 2004.

(Credit: X-Prize)

Fig 8-49 The Ansaris, Burt Rutan, Paul Allen, pilot Brian Binnie and Sir Richard Branson standing in front of SpaceShipOne, after winning the $10M Ansari X-Prize.

(Credit: Don Logan)

Anousheh Ansari became so committed to the whole idea of space tourism that, as we have seen, she subsequently went up with the Russians in a Soyuz orbital mission in September 2006. She also continues to develop her own space tourism company. Anousheh talked in 2007 at a space conference in Arlington, Virginia, and her zeal was obvious when she said: "The human race needs a backup plan!", and "We need a bigger spacecraft than Soyuz to provide for all the space tourists' needs."

The record-breaking SpaceShipOne flew into space, and into the history books, three times in all, and the three flights took place in 2004. We may recall that the 1903 Wright Flyer only flew four times. The first flight of SpaceShipOne into space in June 2004, and the Ansari X-Prize flights, in September and October of 2004, brought thousands of interested public in the early morning to the middle of nowhere in the Mojave desert of California. It was an awesome experience to be on a narrow road, as part of a nose-to-tail procession of vehicles, headlights visible for miles, at that early hour driving across the desert under the stars, and knowing that something extraordinary was about to happen. Even the parking passes printed for the occasion captured the mood (Fig 8-50). As the dawn progressed, and the White Knight carrier plane took off with SpaceShipOne slung underneath, and circled slowly for an hour to gain altitude, the crowd strained to see what was taking place (Fig 8-51). The early morning take-off was chosen for the relatively calm and stable air over the desert at that time of day. Stu Witt's small and disciplined team had made sure that everything was ready for the attempt above his airport.

Fig 8-50 The author's parking pass for viewing the SpaceShipOne flight, June 2004, from the Mojave Spaceport.

(Credit: Author)

Fig 8-51 The crowd at Mojave in 2004 looks towards SpaceShipOne returning from space. They would all go themselves, if they could.

(Credit: Richard Seaman.com)

Amongst that crowd were a number of key players. We have mentioned earlier that Konrad Dannenberg was there. And that Scott Crossfield was there. Now we see in Fig 8-52 Buzz Aldrin to the right, and Sir Richard Branson of the Virgin Group at center, with his Virgin Galactic CEO Edinburgh-born Will Whitehorn to his right.

Lindbergh had himself taken part in that earlier prize competition so many years before. His grandfather had won the Orteig Prize and thereby set in motion the events that led to the aviation and airline business we know and use today. What would be the legacy of the Ansari X-Prize when contemplated with 77 years of hindsight?

Fig 8-52 Sir Richard Branson (center), his Virgin team, and Buzz Aldrin (right) at Mojave for the June 21 flight of SpaceShipOne in 2004.

(Credit: BruceDamer)

Fig 8-53 Burt Rutan and Sir Richard Branson at the Ansari X-Prize events at Mojave in October 2004, with SpaceShipOne. Branson would go on to create a commercial spaceflight operation using Rutan's SpaceShipOne technology.

(Credit: Robert Galbraith/Reuters)

Fig 8-53 registers Burt Rutan with Richard Branson together at the Ansari X-Prize events. We have followed Rutan's work towards reaching this point in earlier chapters. We learned how, even before the Moon landings, he met von Braun and was impressed with his extraordinary vision and charismatic drive. He, in turn, had now inspired great things from a loyal team of supporters. But this photograph captures a moment when the baton of space tourism was metaphorically passed from hand to hand to another charismatic leader. Branson will take the vision onwards.

Let us however linger to note at this moment of success Fig 8-54, which tracks down Lindbergh's grandson, one of the founders and trustees of the competition. He is observing the Ansari X-Prize events at Mojave, and, as an important link in our story bringing it full circle, recalling how Charles

Fig 8-54 Charles Lindbergh's grandson, Erik, one of the founders of the Ansari X-Prize, witnessing the events at Mojave in 2004.

(Credit: Author)

Name Mike Winston Melvill

(Credit: Virgin Galactic)

Summary Description
Rocket plane test pilot
First civilian pilot astronaut, SpaceShipOne, June 21 2004

Date of Birth
November 11 1940, Johannesburg, South Africa

Date of Death
n/a

Nationality South African/US

Achievements
Flew 2 of the 3 flights into space of SpaceShipOne, including the first ever (June 21, 2004), and the first X-Prize flight (September 29, 2004).

Specific help for Your Ticket to Space
Melvill demonstrated that sub-orbital space tourism was possible by flying the SpaceShipOne rocket plane above the internationally recognized 100km limit defining the beginning of space. Twice. Melvill was 64 when he did this. On arriving at the zero-g part of his parabolic trajectory on his first space venture, Melvill released a pocketful of M&M candies to float free and demonstrate the environment for the on-board cameras. Even though the initials of his name look suspicious, he claims no links to the candy firm whose product was thus given prime-time free advertising.

Name William Brian Binnie

(Credit: scaled.com)

Summary Description
Rocket plane test pilot, SpaceShipOne

Date of Birth
1953, West Lafayette, Indiana, USA

Date of Death
n/a

Nationality US (UK father)

Achievements
US naval aviator and test pilot. Flew the SpaceShipOne on its first powered flight (on Wright Brothers' Centenary of December 17 2003), and its second, and winning Ansari X-Prize flight into space, October 4 2004.

Specific help for Your Ticket to Space
Binnie was a test pilot, in 1999, for another attempt at commercial private spaceflight, called Roton, before he joined Burt Rutan's company and became a SpaceShipOne test pilot. Although a US citizen, he is also claimed by Scotland as its first astronaut, since his father was Scottish and he spent much of his childhood in Aberdeen. His first flight of SpaceShipOne under rocket motor power was very risky. He was able to get the spacecraft *ascending* through the speed of sound on this first powered flight, whereas for the record breaking Bell X-1 in 1947, it had been necessary for Chuck Yeager to fly *downwards* to break the sound barrier. The flight took place on the occasion of the Centenary of the Wright Brothers' first flight. His dedication to flight testing space tourism vehicles made your ticket to space a possibility.

Fig 8-55 and Fig 8-56 celebrate a victorious pilot Mike Melvill standing on his craft before the crowds, having successfully returned from space moments before. Someone had handed him a sign that underlined the private nature of the achievement (while perhaps being a little unkind to Patti Grace-Smith and her band of FAA-AST regulators who had underwritten the event!). The sign says "SpaceShipOne – Government Zero". The author talked with Melvill, in August 2005 in New Jersey, and mentioned that the flight had seemed to go dangerously wrong towards apogee. The craft got into a rapid spin and he had continued under rocket power regardless: "I must say I was terrified watching the control problem towards the top of your climb via the big-screen TV". "I was terrified myself", said Melvill. Both Mike Melvill, who flew the June 2004 flight and the first Ansari X-Prize flight in September 2004, and Brian Binnie, who flew the second Ansari X-Prize flight 6 days later on October 4th, received astronaut wings from Patti Grace-Smith of the Federal Aviation Administration. We already gave Mike Melvill a "Wright Stuff" award in the X-Men category in Chapter 5. He flew two of SpaceShipOne's three trips into space, forging the way for the space tourism pilots of the future to follow. He was so determined on success that he kept his rocket engine burning even when his position was perilous, and earned his success the hard way.

Fig 8-56 Mike Melvill returns from space as the first civilian pilot astronaut, June 21st, 2004. The sign reads: "SpaceShipOne - Government Zero".

(Credit: Author)

SpaceShipOne has now completed its work and has ended up in the National Air and Space Museum in the Mall at Washington DC (Fig 8-57). It is in the main hall, and shares pride of place with Chuck Yeager's Bell X-1 and the Ryan Monoplane "Spirit of St Louis" which took Lindbergh from New York to Paris in 1927. It is also not far from another Burt Rutan record breaker, the Voyager. The Wright Flyer of 1903 is nearby.

Fig 8-55 Mike Melvill, first civilian pilot astronaut, is cheered by the crowd on returning from his sub-orbital space trip in SpaceShipOne, Mojave, CA, June 21st, 2004.

(Credit: Author)

Fig 8-57 SpaceShipOne amongst other famous craft at the National Air and Space Museum (NASM), Washington DC. Burt Rutan and Paul Allen attend ceremony when the SpaceShipOne is installed between Lindbergh's "Spirit of St Louis" and Chuck Yeager's Bell X-1.

(Credit: Eric Long/NASM)

Following Rutan's demonstration of the technology of his White Knight/SpaceShipOne combo at the Ansari X-Prize, Branson decided to back the space tourism concept by creating a space tourism business operating under the name Virgin Galactic. He bought Rutan's technology, and set in motion the first fully financed venture to bring space tourism to the public. He subsequently also backed the development of a new spaceport in New Mexico called Spaceport America to be used as his eventual base of operations. Spaceport America had been created largely under the guidance and enthusiastic support of New Mexico's governor at the time, Bill Richardson (Fig 8-58, Fig 8-59).

Fig **8-58** Bill Richardson, Governor of spaceport state New Mexico, with Sir Richard Branson celebrating the signing of the agreement of Virgin Galactic to use Spaceport America as its center of operations.

(Credit: Spaceport America)

Fig **8-59** Architect's projected inside view of the terminal building at Spaceport America, in New Mexico.

(Credit: Spaceport America/Foster & partners)

Name   Sir Richard Charles Nicholas Branson

(Credit: Virgin Galactic)

Summary Description
Adventurer, Airline Owner, Space Tourism business venturer

Date of Birth
18 July 1950, Blackheath, London, UK

Date of Death
n/a

Nationality UK

Achievements
Created and managed Virgin empire of over 350 companies (records, airlines, railways, etc).
Created Virgin Galactic to provide sub-orbital space tourism experiences.

Specific help for Your Ticket to Space
Probably best known in the USA through his television program "The Rebel Billionaire", Branson is a charismatic adventurer and risk taker, as well as being a master of marketing and PR. Branson's spaceline will fly you to space on SpaceShipTwo. What a great name he chose – Virgin Galactic! Branson has always been a rebel, and demonstrated repeatedly to more conventional business players that his approach can be very successful for all concerned. Branson has supported the creation of Spaceport America, the first purpose built space tourism spaceport, by basing his company there, and so becoming its anchor tenant. Richard Branson's decision to risk his brand name in the space tourism business is probably the industry's best guarantee of safe operations.

For space tourism to be a success, it is important that the public does not have to have the health conditioning of a professional astronaut/ fighter pilot. At the time of writing, work is still being carried out to determine what minimum medical requirements will need to be met in order to be eligible for a space tourism flight. For this reason it is fortunate that the Mercury astronaut, and first American in orbit, John Glenn flew on a Space Shuttle in October 1998 when he was 77 years old. Although clearly he was a former professional astronaut, this nevertheless provided some good data to help calibrate the training regime for the modern space tourist. John Glenn nowadays often jokes about his age. He had been the oldest of the Mercury astronauts and used to endure much from his colleagues on that point. When Glenn had his Shuttle flight, Schirra (who had flown in Mercury, Gemini and Apollo spacecraft) said that he too would like to go on the Space Shuttle, "but he was not old enough!". In May 2007, at the Smithsonian in Washington, Glenn gave this offering related to a possible extravehicular activity or spacewalk: "There is no truth in the rumor that they would not let me do an EVA on my Shuttle flight in case I just wandered off somewhere!"

Space tourists in general will not have the same health profile as government astronauts. Generally, to become wealthy enough to afford a ticket to space will take most of us a working lifetime, so notwithstanding the relatively young age of some of the first space tourists, one would expect that space tourists will on average be biased towards their 50's, 60's or 70's. If you are younger, and maybe not as wealthy, as the typical space tourist, you could always work at a spaceport, or maybe hold out for a standby seat. Some further good evidence about the range of conditions acceptable for public spaceflight came when Peter Diamandis made arrangements to fly the physicist Stephen Hawking, who has advanced Motor Neuron Disease, on one of the zero-g flights that his company Zero-G Corp offers. The photo (Fig 8-60) demonstrates that the trip was a success.

Burt Rutan

Fig 8-60 Physicist Stephen Hawking enjoys weightless training in preparation for possible space flight, April 2007, as Peter Diamandis assists.
(Credit: Zero-G Corp/Jim Campbell)

Stephen Hawking is the brilliant Cambridge professor of cosmology whose "A Brief History of Time" explaining the mathematics of black holes became a best-seller in 1988. Hawking gave a lecture in London in April 1992, offering his wry observations about everything from scientists to the human condition. Even his office door in Cambridge carries the sign: "Quiet Please, The Boss is Asleep". His physical limitations have not been allowed to control his effectiveness as a great scientist and communicator. Hawking is a strong advocate of space tourism, and hopes to go himself if possible. His reasons were explained in a 2001 interview with the "Daily Telegraph" newspaper: "I don't think the human race will survive the next thousand years, unless we spread into space. There are too many accidents that can befall life on a single planet. But I'm an optimist. We will reach out to the stars."

Eric Anderson and Peter Diamandis have already won our first "Wright Stuff" awards for the space tourism category, and we here have no hesitation in nominating Messrs Rutan and Branson for our two remaining awards. Rutan showed how a small private operation (about twenty people) could essentially create a private space program and successfully fly a re-usable space plane capable of carrying passengers, against the beliefs of most people in the space field at the time. Branson has very publicly placed his resources, his name, and his company brand, in the service of making space tourism a reality, at a time when most other corporate funders think it is just too risky. He is going forward with an upgraded SpaceShipTwo craft, which will carry more passengers. The new mother-plane White Knight Two was rolled out and unveiled on 28th July, 2008. Branson named it "Eve" after his mother, and she was there to perform the champagne christening. He also commissioned a second copy of the carrier plane, to be named "The Spirit of Steve Fossett", after his old friend and fellow-adventurer. Five SpaceShipTwo craft have been commissioned, with the roll out and unveiling of the first one on 7th December 2009. Test flights of the prototype are to be carried out starting in 2010, with the Virgin Galactic business going operational with the first space tourists, probably sometime in 2012. Branson has stated from the outset, however, that he will only fly "when we are ready" and he will keep testing until that time. He has therefore not stated a definite date for commencement of service.

Sir Richard Branson

Well, the Ansari X-Prize and the "Wright Stuff" prizes have now been awarded. What more of the story remains to be told? Of course, this is a story that is only just beginning. At the time of writing, you and the other first fare-paying passengers have yet to get on board SpaceShipTwo for your ride. It is possible that passengers might fly first on other vehicles than the Virgin Galactic SpaceShipTwo. Because there were others at Mojave to see those magical events in 2004, and who is to know how far they will be able to take their dreams? Fig 8-61, for example, shows Elon Musk.

Musk was a successful Silicon Valley entrepreneur who created the ubiquitous Paypal software. He then established the company SpaceX in 2002 when he was 31 years old, and has demonstrated how a small entrepreneurial firm can build a rocket (Falcon 1) which can put a payload into orbit. He is building more powerful rockets and intends to launch a human-rated capsule called Dragon (Fig 8-62) into orbit aboard his Falcon 9 vehicle (Fig 8-63). Test flights of the Falcon 9/Dragon are expected to begin in the Spring of 2010, just as this book goes to print, so

we cross our fingers and wish for a good outcome. This is a totally new vehicle, and Musk has warned in advance that some kind of problems might be anticipated on a first test launch. That had certainly been the case in the early years of the space program, when sometimes a new rocket experienced a string of failures before becoming successful. Many commentators in 2010 have however forgotten this history, because it has been so long since a totally new US launch vehicle has been introduced. Once it has begun to demonstrate its reliability statistics, Dragon may well become the means whereby orbital space tourism becomes possible from US shores. If so, we can anticipate that in a later reprint of this book, Elon Musk will be a future recipient of a "Wright Stuff" award!

Fig 8-61 Elon Musk watches the Ansari X-Prize events in September 2004 at Mojave.

(Credit: Author)

Fig 8-62 Elon Musk with his Dragon craft that aims at orbital space tourism.

(Credit: SpaceX)

Name Elon Musk

(Credit: Author)

Summary Description
Commercial Space Entrepreneur

Date of Birth
28 June 1971, Pretoria, South Africa

Date of Death
n/a

Nationality US/South Africa

Achievements
Created Paypal Internet software company.
Created SpaceX , first private company to launch satellite.
Owner of Tesla, company producing electric cars.

Specific help for Your Ticket to Space
Designing the Dragon spacecraft, a potential orbital space tourism craft. Elon Musk, who made his money in Internet ventures, has put his personal fortune at risk through creating Space Exploration, Inc (SpaceX) with the express purpose of building a reusable technology for getting people into orbit at a fraction of the cost of previous methods. Musk brings a youthful, and highly motivated, management team to the task. He has grand visions and does not intend to end with only Low Earth Orbit. His company was willing to offer NASA a fixed price proposal for launching cargo and people to the International Space Station, and the proposal had performance milestones that had to be met before he would be paid. So he is no stranger to risk-taking. After 2011, when the Space Shuttle retires, the SpaceX Dragon craft may be the only US spacecraft taking US astronauts up to the space station. And the good news is that there may well be lots of spare seats for orbital space tourists!

Fig 8-63 Elon Musk with his Falcon 9 SpaceX vehicle in 2009 at Kennedy Space Center.

(Credit: SpaceX)

In Figure 8-64 we are at a space tourism award ceremony, and two more enablers of space tourism appear. Robert Bigelow created the hotel chain Budget Suites and is the owner of Bigelow Aerospace. He has designed, and already launched, one-third scale prototypes of a space hotel for tourists. He is building the destinations for Musk's future orbital space tourists, and believes that the habitats may also be used as national or corporate space stations for organizations or nations who for various reasons do not wish to use the ISS. Alongside Bigelow (at right) in the photograph is Rick Searfoss, a former NASA Shuttle astronaut, who is the test pilot for XCOR, which occupies a hangar next to Rutan's at Mojave.

Fig 8-64 Space Tourism awards ceremony in Los Angeles in May 2006, with Burt Rutan, the author, Robert Bigelow, Rick Searfoss.

(Credit: Sarah Fisher)

Jeff Greason, the President and CEO of XCOR (Fig 8-65, with Searfoss), has been steadily building a space capability, starting with pumps, motors and tanks, and his first full vehicle space tourism offering will be the Lynx. However, meanwhile he has been building rocket racers for Diamandis's new Rocket Racing League. Greason devoted a great deal of time and detailed commitment to the development of the regulations for space tourism, working with the FAA-AST through various committees. The author also worked with the Commercial Space Transportation Advisory Committee (COMSTAC) of the Federal Aviation Administration to help ensure that the emerging regulations for space tourism were as least intrusive as possible. The process was to send in amendments to draft legislation, often forty-pages thick, via something called the Proposed Rulemaking Procedures. However abstruse or complicated or apparently unimportant was the article under discussion, one could always find Jeff Greason's intellectual fingerprints all over the draft. He was also a Commissioner with the 2009 Augustine Commission, tasked with a Review of US Human Spaceflight Plans for the new Obama Administration, and he was a powerful voice for the inclusion of a commercial human spaceflight component in any long term national human spaceflight plans. He is Vice Chairman of the industry group The Commercial Spaceflight Federation.

Fig 8-65 Jeff Greason with his chief test pilot Rick Searfoss at March 2008 press conference in Los Angeles, California.

(Credit: XCOR)

Figure 8-66 places Searfoss with one of the XCOR rocket racers. Peter Diamandis invented the Rocket Racing League as a way to sustain public interest in rocketry and public space travel

Name Robert T Bigelow

(Credit: Space News photo by Barbara David)

Summary Description
Hotel business entrepreneur, manufacturer of space hotels.

Date of Birth
1945, Las Vegas, Nevada, USA

Date of Death
n/a

Nationality US

Achievements
Created Budget Suites of America hotel chain.
Founded Bigelow Aerospace and designed range of space hotels and launched two prototypes.
Created $50M America's Space Prize to encourage private flights into orbit.

Specific help for Your Ticket to Space
We can thank the somewhat reclusive Robert Bigelow for the fact that orbital travelers will have a non-governmental destination, specifically designed for space tourists, rather than a governmental space station to visit in orbit. Bigelow launched the "Genesis 1" prototype hotel in July 2006. The "Genesis 2" followed a year later. These spacecraft are built using an inflatable technology originally developed at NASA. He is ready to launch a full-scale version, called "Sundancer", and has the door-keys ready for the guests, but is waiting for someone to develop the technology to bring his customers to the door of their orbital destinations. Bigelow has attempted to prime the pump towards solving this rather awkward delivery problem, but his America's Space Prize has not produced the solution. Who will deliver his first customers to his orbital hotels? It may well be Elon Musk's SpaceX, with their Falcon vehicle and Dragon spacecraft, which could take 6 visitors at a time to their orbiting "Budget Suites" vacation hotel.

Name Jeff Greason

(Credit: XCOR Aerospace)

Summary Description
Commercial space tourism entrepreneur

Date of Birth
1967, USA

Date of Death
n/a

Nationality US

Achievements
Team Lead at Rotary Rocket.
CEO of XCOR, designer of Lynx vehicle.
Designer of rocketplanes for Rocket Racing League.
Member of 2009 Augustine Commission into future of human spaceflight in US.

Specific help for Your Ticket to Space
Jeff is a methodical and careful aircraft designer who has created probably the lowest price offering for sub-orbital space tourism, with his Lynx vehicle. Lynx is the minimalist's approach to space tourism. The craft is so small that there is only room for one passenger. The passenger must sit next to the pilot for the trip, and thus experiences the identical views. The minimalism extends to the fact that the Lynx will not initially quite reach the officially recognized "boundary" of space when it arrives at its apogee at around 60 km. The best part of the minimalism is the price, which will be only half of the initial prices announced by Virgin Galactic for their offering. His XCOR operation is in the hangar next door along the flight-line at Mojave Spaceport from Burt Rutan's Scaled Composites, where the Virgin Galactic craft is being built.

on an ongoing basis even before the full space tourism flights were to become available. The idea was to copy the NASCAR motor racing event's level of interest and advertising support. Unlike the NASCAR's, however, these rocket racers will have to compete in a racetrack in the sky at a contest at each year's X-Prize Cup events: 3D instead of 2D racing. Developing the vehicles for the League would also help develop rocket technology, and the operational readiness skills that would be necessary for regular sub-orbital space flight operations.

Fig 8-66 Rick Searfoss and XCOR Rocket Racer.
(Credit: XCOR)

It seems that the challenges of the new space tourism industry have attracted several ex-NASA astronauts. Jim Voss (Fig 8-67) is another former Shuttle Commander now engaged in space tourism developments. In Fig 8-68, we observe him with Buzz Aldrin inside a full-scale engineering mockup of the CXV built by Transformational Space, Inc. (t/Space), for which Voss was then the chief engineer. The t/Space company was operated by David Gump (Fig 8-69) and serial space entrepreneur Gary Hudson (Fig 8-70), both long-standing space tourism advocates. Their business plan, in which the author assisted in business development and procurement matters, was to develop a spacecraft, called the CXV, by means of a fixed price contract with specified performance-related funding milestones. This would be used by NASA to deliver astronauts and supplies to the International Space Station (ISS), and it would have spare seats to provide for orbital space tourism participants.

Their ideas were eventually adopted by NASA, and transformed into the so-called NASA Johnson Spaceflight Center's COTS (Commercial Orbital Transportation Services) program, although t/Space itself did not in the event win the competition to receive the funding to proceed. The main beneficiaries of the program have been SpaceX and the more traditional space manufacturer Orbital, both of whom are trying to develop the new vehicle that the US needs, once the Space Shuttle is retired, to get astronauts and supplies into orbit. Orbital space tourism from a US base will be an almost incidental beneficiary of the program. Once either the SpaceX Dragon, looking like the t/Space CXV or the Orbital Cygnus, is operational, then the US will not be dependent on Russian Soyuz transports for either governmental astronauts, or orbital space tourists.

Fig 8-68 Jim Voss (left) gives Buzz Aldrin a tour of a full-scale engineering mockup of his new orbital space vehicle, the t/Space CXV, at a Washington DC space conference in 2005.

(Credit: Author)

Fig 8-69 David Gump with full-scale engineering mockup of the t/Space CXV orbital space tourism vehicle in 2005, a precursor of the SpaceX Dragon.

(Credit: Author)

Fig 8-67 Former Shuttle Commander Jim Voss is now engaged in space tourism developments.

(Credit: NASA)

Fig 8-70 Gary Hudson, of Roton, t/Space and Air Launch, explains his latest commercial space transportation vehicle proposition.

(Credit: Jeff Foust/The Space Review)

Fig 8-71 lines up the FAA's Assistant Administrator (at that time) Patti Grace-Smith with a selection of space tourism astronauts , including Anousheh Ansari, and SpaceShipOne's Brian Binnie, at a February 2007 FAA-AST space tourism conference. The annual Washington DC FAA Space Conferences had focused the community on the progress being made on regulatory matters. This progress was reflected as more and more government astronauts began to leave NASA and join the new entrepreneurial space tourism companies, where they knew they would have more responsibility, more autonomy, and quite frankly more flights into space.

It is interesting to note in the grouping of Fig 8-71 that there are three different kinds of astronauts represented, and in the future as the industry develops, there might be a need to recognize this by some modifications to the definition of "astronaut". Brian Binnie on the right is an astronaut by reason of his having piloted a spacecraft (SpaceShipOne) above 100km. Gibson and Herrington and Thuot have all flown as astronauts on the Space Shuttle (and hope to eventually be flying tourists into space when the new space tourism operators develop the craft). Ansari is an astronaut because she, too, has been into space (in fact into orbit) and yet as a space tourist she was not strictly in command of the space vehicle. Other ex-NASA astronauts who have been supportive to space tourism include Byron Lichtenberg, who helped Diamandis work out the rules for the X-Prize, Frank Culbertson, who has labored behind the scenes to develop the training requirements for the crews and passengers of this new industry, and Ken Bowersox, who is responsible for mission assurance at SpaceX.

In Fig 8-72 we are at a 2006 space conference and see Buzz Aldrin with Robert Bigelow, whom we met earlier, discussing Bigelow's progress in designing, building prototypes, and launching space hotels. Bigelow uses an inflatable technique which had been developed initially by NASA in the late 1990's and which they called Transhab. He has acquired the patents to Transhab and so far launched two sub-scale prototypes of his space hotels in 2006, 2007 as proof of concept. Bigelow is based in Las Vegas, Nevada, and uses Russian launch capability because of its low

cost and convenience. Fig 8-73 records the first of Bigelow's Genesis prototypes in orbit, above Baja California. His main problem now is how to get his guests to the front door in orbit. He has created the $50M America's Space Prize, for the first non-government entity to put people in orbit, but unfortunately (or fortunately, depending on your viewpoint) his money remains safe.

Fig 8-71 A lineup of various "space tourism astronauts" with Patti Grace-Smith, the then US Government regulator (Hoot Gibson of SpaceDev [now Sierra Nevada Corp], John Herrington, then of Rocketplane, Anousheh Ansari of Prodea, Patti Grace-Smith of FAA-AST, Pierre Thuot of the European Space Agency, Brian Binnie of Scaled Composites).

(Credit: Author)

Bigelow was not incidentally the first entrepreneur who aimed at providing private habitats in space. In the early 1980's Bob and Rick Citron raised $100M in investments, built Spacehab, and have had their modules flown via the Space Shuttle. Spacehab was not intended to be a free-flyer, however, and remained in the Shuttle bay during its operations. In April 1998 the author presented a paper: "Station Mir – The First Hotel for Space Tourists?" at the Space 98 Conference in Albuquerque, New Mexico. The idea in the paper was to preserve the aging Russian Mir space station in orbit until such time that an orbital space tourism business could take advantage of the facility. In early 2000, Walt Anderson (who provided the funding for the Space Frontier Foundation) Jeff Manber and co-investors turned the idea into reality and created MirCorp, negotiating with the Russians to lease their station for commercial use. The space station, at Mircorp expense, was boosted to a higher orbit to preserve the options for commercial use, but unfortunately Mir was de-orbited for political reasons in 2001 when the ISS began to be assembled. All of that work in Russia was not totally lost, however, because Mircorp was able to assist in Dennis Tito's attempts to get a ride.

Fig 8-72 Buzz Aldrin (left) discussing space hotels at 2006 space conference with Robert Bigelow.

(Credit: Jeff Foust/The Space Review)

Fig 8-73 Bigelow Aerospace's Genesis prototype inflatable hotel   in orbit over Baja California (2006). This is the model for future commercial space travel destinations.

(Credit: Bigelow Aerospace.com)

We have reached the present day, at the end of our journey through a century of effort and courage and achievement and dreaming and risk taking. Orbital space tourism already operates in a limited way using spare seats on the Russian Soyuz vehicles. SpaceX (as a winner of the NASA COTS competition) is working at increasing the availability of orbital space tourism seats through its Dragon spacecraft, which will operate from US spaceports. Maybe the Orbital Cygnus, another winner of a COTS contract, will also succeed in providing orbital tourism seats from US spaceports. Sub-orbital space tourism will have its first paying passengers very soon, you included, but we still do not know which operator will be the first, or indeed the one you will have chosen with which to fly into space. You do have options. The next few images show some of the contenders, each of which is offering a different design approach to providing the future customers with their ride of a lifetime. Of course, not all of them will necessarily succeed in reaching the point when they can offer services. You should take note of the fact that some of them have already produced hardware, and some have even flight-tested the hardware, whereas many remain at the concept stage. Nowadays computer technology can produce images that seem very lifelike, so be careful to note the difference between real hardware and simulations.

Fig 8-74 (and the cover) show the Virgin Galactic offering; the very real SpaceShipTwo craft, that will be launched from the underside of the WhiteKnightTwo mother plane, which is already well on the way through its flight test regime. Virgin Galactic has stated that it will do its testing of SpaceShipTwo and maybe offer its first services from Mojave, then convert to Spaceport America in New Mexico when that spaceport is ready. Initial premium prices are stated to be $200,000 per seat.

Fig 8-74 Virgin Galactic's SpaceShipTwo, the VSS "Enterprise", suspended under WhiteKnightTwo in its Mojave hangar, December 2009.

(Credit: Virgin Galactic/EPA/Ned RocknRoll)

Rocketplane Global is developing a completely different architectural solution for sub-orbital space tourism, and their Rocketplane offering is shown in Fig 8-75. A regular speaker at commercial space conferences throughout the first decade of the 21st century has been their director of business development, Chuck Lauer (Fig 8-76). The Rocketplane vehicle would take off and land using turbojet engines, and convert to rocket power when at altitude. Rocketplane Global has intended to operate from Oklahoma Spaceport, which is a former Strategic Air Command airfield that is being converted for space tourism purposes. It has received authorization from the FAA's Office of Space Commercialization to operate in this way, but unfortunately the company has experienced difficulties which could delay its entry into operation. One always thinks, however, when looking at Chuck Lauer, that there is a new plan afoot to save the company and its space plane offering, which Chuck, or Rocketplane's financier George French, will reveal, but not just yet.

Fig 8-75 Simulation of the Rocketplane craft at the edge of space.

(Credit: Rocketplane Global)

A third architecture is being developed, as we have seen, by XCOR, and they intend to fly their Lynx rocketplane all the way into space on rocket power, starting with a horizontal takeoff, and landing as a glider (Fig 8-77). On one occasion in 1949 Chuck Yeager flew the Bell X-1 from a standing start on rocket power, so the approach has been demonstrated before. Lynx will only carry one passenger at a time, and will operate from Mojave. This will provide its one passenger with a very "hands-on" astronaut experience, sitting next to the pilot as he executes all the flight maneuvers. The Lynx prototype vehicle will not initially be able to fully reach the "100km"

definition of the boundary of space, probably reaching around 60km, but will nevertheless go high enough to leave the atmosphere behind. The Mark II production version will exceed the 100km limit, however. XCOR is going to charge prices of $100,000 per passenger, thus much lower than Virgin Galactic, and is planning up to 4 sorties per day.

Fig 8-76 Chuck Lauer, Rocketplane's Business Development point person.

(Credit: Rocketplane Global)

Fig 8-77 Simulation of XCOR Lynx sub-orbital rocketplane.

(Credit: XCOR)

We have already seen some of the test flights of the Armadillo vehicle at X-Prize Cup events. It is the stated intention of the company to offer a totally different space tourism experience that will involve a straight up and down vertical trajectory, and Fig 8-78 presents an early concept illustration of this experience. The spaceport has not yet been chosen. David Masten's space systems company also has plans to provide a succession of capabilities for low cost access to space, reaching from unmanned sub-orbital cargo transportation all the way to crewed orbital vehicles, building on his 2009 successes in the Northrop Grumman Lunar Lander X-Prize Challenge.

Fig 8-78 Simulation of the Armadillo "Six-Pack" space tourism concept.

(Credit: Armadillo Aerospace)

Another offering that plans to operate with vertical take-off and landing is being developed by the Blue Origin Corporation, owned and managed by Jeff Bezos (Fig 8-79), who created Amazon.com. Test flights of a prototype, called "Goddard", have been carried out from their private Texas spaceport in Culberson County. The operational vehicle will be called "New Shepard" to recognize the fact that Alan Shepard, the first American astronaut, flew his first mission with a similar trajectory going into sub-orbital space and landing within a 15 minute flight duration. Fig 8-80 showcases the prototype vehicle that has already carried out several test flights. Furthermore, Blue Origin received some NASA funding ($3.7M) via a Space Act Agreement that was awarded in February 2010. NASA views the company as one of the potential commercial providers of space access required under the new direction for the agency announced by its new Administrator Bolden on February 1st, 2010. It seems that Blue Origin may be aiming at orbital destinations after "New Shepard" has demonstrated the sub-orbital capability.

Name Jeffrey Preston Bezos

(Credit: Blue Origin)

Summary Description
Space tourism entrepreneur

Date of Birth
January 12 1964, Albuquerque, New Mexico, USA

Date of Death
n/a

Nationality US

Achievements
Created the Amazon.com web-based company.
Created Blue Origin to build and fly space tourism craft.

Specific help for Your Ticket to Space
For a developer of such an ubiquitous Internet business as Amazon, Bezos has rather surprisingly chosen to develop most of his space program elements, such as his vehicle itself, in extreme secrecy. One cannot help but wonder at the possible reason for this secret space tourism program. Who precisely is he hiding from? It cannot be the Government, because the FAA has licensed all his test launches. It cannot be the future space tourists, because he needs them to sign up and pay their deposits. It must be his competitors, almost all of whom were his competitors once before, when he was building an Internet business to create his first fortune. Nevertheless, you could be flying into space via Amazon pretty soon.

Fig 8-79 Jeff Bezos, President and CEO of the Blue Origin space tourism company.

(Credit: Niall Kennedy)

Fig 8-80 Blue Origin's "Goddard", the prototype for the "New Shepard" sub-orbital space tourism vehicle, returns to its hangar after a flight test. The flight tests, and even the spacecraft and the hangar have so far been kept secret from the public.

(Credit: Blue Origin)

Another potential solution is the Excalibur-Almaz offering (Fig 8-81). This would be a provider of orbital space tourism experiences, using Russian hardware that was developed during the Cold War, with Western financing, and a headquarters in the tax-exempt Isle of Man in the Irish Sea between Britain and Ireland. The company has already acquired several capsules from the former Russian manufacturer in preparation for offering services.

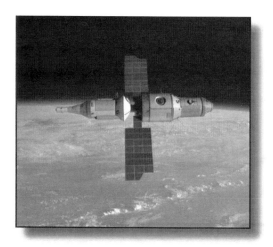

Fig 8-81 Simulation of the Excalibur Almaz concept for orbital space tourism using Russian hardware.

(Credit: Excaliburalmaz.com)

The Sierra Nevada Corporation, maker of the hybrid rocket motor for SpaceShipOne and SpaceShipTwo, is developing its own space tourism craft, the "Dream Chaser" (Fig 8-82). This is based upon an earlier winged lifting body spacecraft demonstrator, the NASA HL-20, from a canceled X-plane program. Sierra Nevada has received some $20M of NASA Space Act funding in February 2010, so that NASA can continue to monitor the developments of their discarded design in its new application as a space tourism vehicle. Jim Voss, whom we met earlier helping to design an orbital crew capsule for t/Space, has now with the demise of that company moved on to Sierra Nevada as Director of Advanced Programs, and program executive on "Dream Chaser". Clearly Jim, who is a highly experienced Shuttle astronaut, has decided that the future is in entrepreneurial space ventures. We can expect to continue to see him doing whatever it takes, wherever he can help, to get us all into space where we can also share in his spaceflight perspectives.

The array of companies involved in space tourism shows just how dynamic the field is. Some of the companies are announced with great plans and a fanfare, then nothing emerges even years later. The author has tried to offer the reader a focus on companies that have probably the best chance of success. You will have no doubt noticed that not all of these companies so far described are of US origin. Virgin Galactic, which is perhaps the leading contender, has British heritage. There are other British firms with perhaps fewer prospects of success (e.g. Steve Bennett's Starchaser and David Ashford's Bristol Spaceplanes) which have therefore not been described in detail. There is a firm with partial Canadian and partial US/Indian heritage, PlanetSpace, which likewise does not at the time of writing seem to be able to demonstrate much progress, at least publicly, towards offering space tourism tickets. The major European aerospace manufacturer EADS indicated at one time its interest in developing a space tourism vehicle, but has so far not advanced beyond its initial outlines. A Swiss-German company Talis Enterprise is also proposing a winged spaceplane called Black Sky, with rides which will be priced below $50,000 per passenger, but which would not reach as high as even XCOR's Lynx vehicle. Only the US has so far put in place a regulatory regime where sub-orbital space tourism vehicles can operate, so foreign space tourism vehicles would have to operate from the US. Space tourism is such a bold reflection of American values. Most of the firms described in this chapter are US companies, and this is partly due to the entrepreneurial spirit that flourishes in the US. There has been a tradition of "new space" in the US that has been striving for over two decades to reach this point of having real space hardware ready to fly commercial passengers, as opposed to simply flying US government astronauts on US government launch vehicles.

During 2004, the advocates of the emerging space tourism industry were excited to follow the Ansari X-Prize competition and to wonder who could or would win the $10M prize for what at the time seemed an incredibly difficult thing to do. As we now know, Burt Rutan, with his sponsor Paul Allen, won and thereby opened up the possibility of the sub-orbital space tourism opportunity for the many who wanted to undertake the experience. Now there is a new race taking place, with even bigger stakes. The race this time is about the provision of commercial space tourism flights on a regular basis. Who will be first? Who will be most successful? Which of the varied architectures will the space tourists prefer? With which company will you have your flight? We do not have long to wait to learn the answers. Although of course there needs to be a focus on safety to be achieved

through testing, and so the race this time is about who can be first in achieving the requisite safety testing in order to become operational.

Fig 8-82 A simulation of Sierra Nevada's "Dream Chaser" space tourism concept craft.

(Credit: Sierra Nevada Corp)

There are also ancillary businesses that have emerged to support the space tourism sector. Some businesses offer training for both crew and passengers, or medical services and screening (e.g. Vernon McDonald's Wyle Labs operation). Dr Harvey Wichman and his students from Claremont Mckenna College in California have carried out many practical experiments to study psychology in the space tourism context. One company, as we have seen, offers space hotel facilities. Some offer space insurance, or investment financing (where Paul Eckert has made a steady contribution), or consultancy.

Sensing the long-awaited change in the times, a long-time commercial space activist, and a founder of the Space Frontier Foundation, Rick Tumlinson (Fig 8-83), has put away his pirate's eyepatch, become respectable, and created the firm Orbital Outfitters with Jeff Feige to provide spacesuits for space tourism crew and passengers. And there are the activities of the builders and operators at each of the new spaceports. There are also specialized travel agents like Eric Anderson's

Space Adventures and Jane Reifert's Incredible Adventures, which will bring in the customers. An industry association, the Commercial Spaceflight Federation, has even been formed, with Brett Alexander at its head, to represent the common interests of future space tourism industry companies on Capitol Hill. Alexander, who had more than a little involvement in the writing of George Bush 43's Vision for Space Exploration, is now, in the Obama regime, chairing the Commercial Space Committee to the NASA Advisory Council. He is a consummate "Beltway Insider", who adroitly moves from one key job to another even as presidential administrations change from one party to another. Alexander clearly finds the political process itself at least as fascinating as commercial space and space tourism, and it's just as well for the emergence of the space tourism industry that *somebody* does! Because the emerging industry is inevitably going to be in the spotlight on the Hill during the transition towards commercial spaceflight and away from everything for human spaceflight in the US having to be NASA-built.

Fig 8-83 Rick Tumlinson, a commercial space activist, now founder of Orbital Outfitters.

(Credit: Orbital Outfitters)

The future, of course, ultimately belongs to our children. They will inherit the new world of possibilities described in this book. A number of educators have emerged who are leading efforts to inform the new generation about space commerce and space tourism, including Nicholas Eftimiades with his Federation of Galaxy Explorers, the Challenger Learning Centers, Lonnie Schorer with her "Kids to Space" operation, and Elizabeth Wallace and her Starry Telling Festivals. Fig 8-84

takes place in the then-empty desert at Spaceport America's New Mexico location before building has commenced; empty, that is, except for scrub and tumbleweed and lizards. Sir Richard Branson, who amongst his many attributes is a master showman, is shown with a group of children firing off models of SpaceShipOne, to celebrate the agreement to build the spaceport.

Fig 8-84 Sir Richard Branson with future space tourists from East Picacho Elementary in Las Cruces, at the site of the future Spaceport America in New Mexico, in 2004.

(Credit: Spaceport America)

Fig 8-85 illustrates the winning architect Sir Norman Foster's design for the completed Spaceport America, which blends in with the beautiful scenery of the New Mexico desert. As the book goes to print, the runway is almost complete.

Space tourism, it is hoped, will reinvigorate the public to the possibilities and promise of space travel, and make it more aware of the fragility of the planet. From space, as you will soon see on your own flight, it becomes all too obvious how thin is the atmosphere that protects us all, and how insignificant are traditional man-made boundaries and frontiers. The re-usable technologies and operational concepts brought about as a result of space tourism will lead to safer, more reliable and less costly ways to get payloads into space for all purposes. Future goals might even include developing new forms of energy, and new sources of resources, for use on this planet in the decades and centuries ahead. New spaceports will be set up which will bring employment, entertainment and education to their visitors.

Fig 8-85 Architect's concept of the future Spaceport America in New Mexico.

(Credit: Spaceport America/ Foster & Partners)

What would von Braun have thought about this Spaceport America facility, not very far from his new home in the US at White Sands? What would Lindbergh have felt if he could have seen it just over the mountain range from the first launch sites of the Goddard rockets? Since von Braun and Lindbergh are both recipients of "Wright Stuff" awards, we must conclude that they would not have been surprised, but that they would have smiled to see how their dreams were being realized.

Fig 8-86 Burt Rutan, Max Faget, Buzz Aldrin, with White Knight and SpaceShipOne, Mojave, CA, April 2003.

(Credit: Jeff Foust/ The Space Review)

A cameo incident had happened, at the April 2003 official unveiling and roll-out of SpaceShipOne in Mojave, when three men met

briefly to consider what was about to take place. We have met two of them already. One was Burt Rutan. Another was Buzz Aldrin. The third member of this select grouping was the late Max Faget, who had designed all three early US NASA space capsules, the Mercury, Gemini and Apollo spacecraft (Fig 8-86). It was Faget who had worked out how to re-enter the atmosphere using ablative techniques. He must have been mightily impressed to see for himself how Burt Rutan's simple device of the feathered wings was about to make sub-orbital space tourism possible.

In Fig 8-87, we are invited to observe a close up of two of the engine cowlings for the Virgin Galactic WhiteNightTwo vehicle "Eve". Each of the engines carries a design which underlines the development heritage of the space tourism craft. In succession starting at the bottom, we see Icarus, then the Wright Flyer, then Lindbergh's "Spirit of St Louis", then Yeager's Bell X-1, then a Boeing jetliner, a Lunar Lander, SpaceShipOne, and finally the WhiteNightTwo/SpaceShipTwo combination.

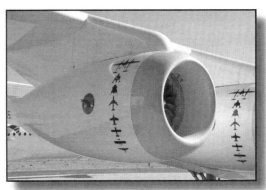

Fig 8-87 WhiteNightTwo, mothercraft for Virgin Galactic's space tourism trips, displays its heritage.

(Credit: Jeff Foust/ The Space Review)

What does the distant future hold for space tourists? Well, some believe that there will be a follow-on to the Concorde which will enter space as it goes from point to point around the Earth. One such person is Alan Bond of Reaction Engines, Ltd (Fig 8-88), who was one of the visionaries at the British Interplanetary Society, mentioned earlier in this history. He has spent his professional life designing the special kind of engine that would be needed for such point-to-point trips, fueled by liquid hydrogen, the ultimate "green fuel". It is even possible that this kind of technology might eventually get passengers and cargo all the way into orbit. Another firm pursuing this long-term goal is Ajay Kothari's Astrox Corporation in Maryland. Also John Old's consulting firm Spaceworks Engineering, Inc. (SEI) based in Georgia is leading the Fast Forward Study Group to explore the practicalities of possible Point-to-Point sub-orbital space travel, including the exciting technologies and the perhaps less-motivating but equally significant hard commercial realities. In India, the agencies are investigating hypersonic possibilities through their Avatar and Hex programs. So far, the jury is still out regarding whether the markets for such hypersonic vehicles are sufficient to justify the development costs, and even to know whether point-to-point markets would be mostly cargo or mostly passengers. However, the technical concepts themselves have an undeniable beauty (Fig 8-89), as of course do the views that space tourists like yourself may expect to soon experience (Fig 8-90).

Fig 8-88 Alan Bond, CEO of Reaction Engines, with a model of his Skylon point-to-point hypersonic sub-orbital space transporter.

(Credit: Reaction Engines)

You picked up this book because you were about to go into space on a space tourism vehicle and you wanted to know more of the background of this industry that was about to make your dream possible. You now know the story of the century of effort behind your ticket to space and of the extraordinary and diverse range of individuals whose work, and often personal courage, has led to this point: all the way back to the Wright Brothers themselves. You also know why space tourism is not just important in its own right, but

is a vital enabler of reusable space technologies that will provide long term benefits for all space uses and users off planet Earth. Hope you enjoyed the ride, and awarding the prizes along the way. It was intended for fun; there will be no test.

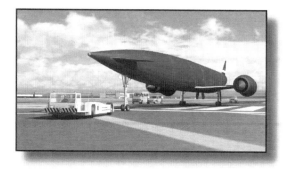

Fig 8-89 The Reaction Engines Skylon – simulation of a concept for a possible future point-to-point hypersonic sub-orbital space vehicle.

(Credit: Reaction Engines)

Fig 8-90 The space tourist's view from space.

(Credit: NASA)

Fig 8-91 and Fig 8-92 remind us of the journey which you have just followed. OK, so you noticed that we relented and gave Buzz an award after all! Looking at the list of "Wright Stuff" awardees, it just did not look right without him. How could we leave him out? You have seen his efforts everywhere throughout this narrative. He is the only member of the NASA astronaut team who has devoted his life since leaving the corps to the creation of commercial spaceflight and making it possible for you to have a ticket to space. He certainly more than meets the specs. Sometimes you just have to break the rules, even your own rules, to do what's right!

Now, it's time for your turn to make your own personal contribution, to have a great trip into space, and then spread the word to your friends. And particularly, if you are still in your twenties or thirties, help this show along as did so many young people during the last century. In the words of Alan Shepard, "Let's light this candle"!

FIG 8-91 SUMMARY OF "WRIGHT STUFF" AWARDS	
CATEGORY	RECIPIENT
AVIATORS	The Wright Brothers
	Charles Lindbergh
ROCKET MEN	Sergei Korolev
	Wernher von Braun
X-MEN	Scott Crossfield
	Mike Melvill
PRESIDENTS	John F Kennedy
	George W Bush (43)
ARTISTS	Chesley Bonestell
TOURISTS	Eric Anderson
	Peter Diamandis
	Burt Rutan
	Richard Branson
	Buzz Aldrin

Buzz Aldrin

Name Dr. Edwin E. "Buzz" Aldrin, Jr

(Credit: Sharespace.com)

Summary Description
Former Moonwalker, space tourism advocate.

Date of Birth
January 20 1930, Montclair, New Jersey, USA

Date of Death
n/a

Nationality US

Achievements
Air Force Colonel with Korean combat service. Gemini astronaut and Lunar Module Pilot of first Moon landing.

Specific help for Your Ticket to Space
Buzz is the son of a father who was a USAF colonel and a mother whose maiden name was Marion Moon. So maybe it's not surprising that he became a pilot, then an astronaut in the Apollo era, and went to the Moon. But what Buzz did *afterwards* is what makes him a major contributor to your ticket to space. Buzz Aldrin has been a relentless promoter of the cause of commercial spaceflight, and space tourism in particular. The term relentless has been chosen carefully. Wherever and whenever there have been discussions and workshops and conferences with commercial space on the agenda, Buzz has been there. His impact has been to gradually wear down any opponents to his vision, because it's the only way they can get him to go away. He has been very concerned with the motivation of today's young folk, especially towards encouraging them to study hard engineering and thenceforth to make their impact on the future of space exploration. When you go to the spaceport to take your own spaceflight, there is a very good chance that Buzz will be around somewhere to cheer you on your way.

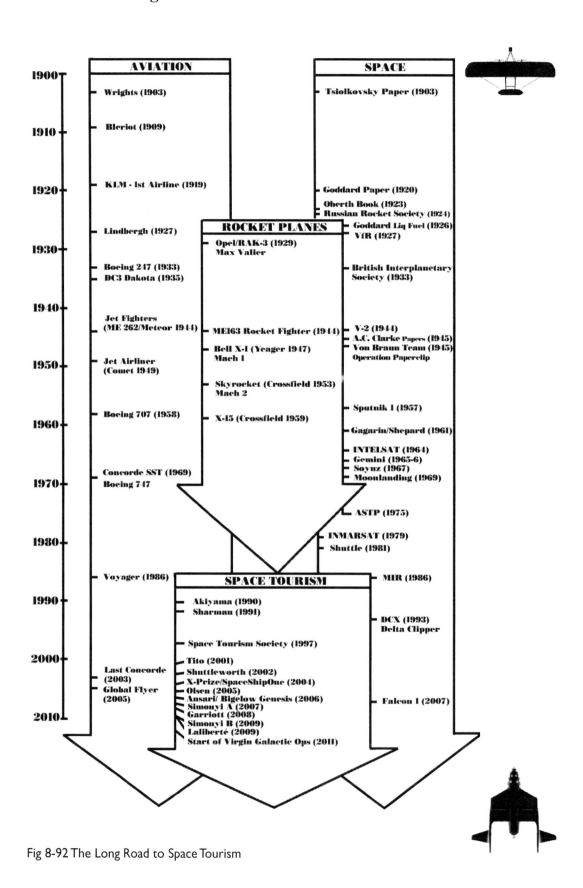

Fig 8-92 The Long Road to Space Tourism

Name Your Name Here

(Credit: Free Clipart)

Summary Description
What will be your contribution to space tourism?

Date of Birth
Confidential

Date of Death
n/a

Nationality Citizen of world

Achievements
How did you get to this point?

Specific help for Your Ticket to Space
What are you going to do to make it possible for others to enjoy a trip into space?

ACKNOWLEDGEMENTS

This book honors and acknowledges all of the pioneers who had the vision and took the risks to ultimately make space tourism possible, so I shall not repeat all of their names here. I look instead to recognize here those whose interactions over the years led to the possibility and realization of the book itself, although regrettably not all are still around to receive that recognition.

The greatest influence on me with regard to aviation was in my early teens joining the Air Training Corps of the RAF, where I enjoyed flying in Chipmunks and Meteors, and where I was able to fly solo in gliders at age 16. Thanks to all the pilot instructors from the RAF and the Northumbria Gliding Club (Waggott, Marples, Stephenson, Ruffell, Wilson) who gave me the training; you showed much courage and perseverance.

With space, I owe my early interest to a wonderful BBC radio program back in 1953 called "Journey into Space", where the hero was a British space pilot long before the word "astronaut" had been invented. Later, I joined the space business myself. At the former Hawker Siddeley Dynamics thanks to Bill Graham who first let me loose on real satellite blueprints, and trusted my post-flight assessments from the Woomera rocket firings. At Inmarsat, thanks to Noel Isotta who gave me the best job in the world – Head of Procurement, buying satellites and launch vehicles for an international satellite services provider. In moving to the US, where I became a citizen in 2000, I thank my many consulting clients (including Messrs Kopp, Gump, Olds, Miller, and the Futron team) who have ensured that I could keep bread on the table. At Tachyon, thanks to Sa'id Mosteshar and to my team in Amsterdam who brought a great satellite broadband service to Europe.

Thanks to Buzz for providing the Foreword, (and to Lisa Cannon for reminding him to do so!). Thanks to Rob Godwin at Apogee for his idea of linking the space tourism story backwards in time to the start of aviation and including the astronaut anecdotes, and for his great cover design. All of the photo sources are acknowledged with each photo. Every effort has been made to trace the copyright holders and identify the correct person or organization for credit. Please advise me if I have in any case given credit incorrectly, so that we can fix it.

Thanks to daughter Grace, who has shared many of her dad's space dreams, and who has given so much to enable this old Brit to become a young American, and to Erik who keeps the website: www.SpaceportAssociates.com afloat. Finally, thanks to my understanding wife, Sarah Fisher, for her uncompromising support. This included reading at least six drafts of the book from start to finish. All omissions and errors remain, of course, my own.

Derek Webber
Bethesda, MD, USA
March 2010
DWspace@aol.com

ACRONYM GLOSSARY

A

AIAA — American Institute of Aeronautics and Astronautics
ATC — Air Traffic Control
ATC — Air Training Corps
ARS — American Rocket Society
ASCENT — Analysis of Space Concepts Enabled by New Transportation
ASTP — Apollo Soyuz Test Project

B

BIS — British Interplanetary Society

C

COMSTAC — Commercial Space Transportation Advisory Committee
COTS — Commercial Orbital Transportation Services
CSF — Commercial Spaceflight Federation
CSLAA — Commercial Space Launch Amendments Act
CXV — Orbital space tourism vehicle for t/Space

D

DC-X — Delta Clipper Experimental single stage to orbit craft

E

EADS — European aerospace company
ELV — Expendable launch vehicle
ESA — European Space Agency
EVA — Extra-Vehicular Activity (spacewalk)

F

FAA — Federal Aviation Authority of US Government
FAA-AST — Commercial spaceflight division of FAA

G

Genesis — Space hotel made by Bigelow Aerospace
GEO — Geostationary orbit (Clarke orbit)
GPS — Global Positioning System

H

HOTOL — Horizontal Takeoff and Landing

I

IAF — International Astronautical Federation
ICBM — Intercontinental Ballistic Missile
Inmarsat — International Mobile Satellite Organization
Intelsat — International Telecommunications Satellite Organization
ISS — International Space Station
ISU — International Space University
ITU — International Telecommunications Union

J

JPL — Jet Propulsion Laboratory (Pasadena)
JSC — Johnson Space Center (Houston)

K

KSC — Kennedy Space Center (Florida)

L

LEM — Lunar Exploration Module
LEO — Low Earth Orbit
LLRV — Lunar Landing Research Vehicle

M

ME — Messerschmitt
MIR — Russian Space Station (it means "Peace")
MIT — Massachusetts Institute of Technology
MSFC — Marshall Space Flight Center (Huntsville)

N

NAC — NASA Advisory Council
NACA — National Advisory Committee for Aeronautics
NASA — National Air and Space Administration
NASM — National Air and Space Museum of the Smithsonian Institute
NSS — National Space Society

P

PtP — Point-to-Point sub-orbital space transportation

Q

QC — Quality Control

R

RAF — Royal Air Force
RLV — Reusable Launch Vehicle
RRL — Rocket Racing League

S

SEDS — Students for Exploration and Development of Space
SpaceX — Space Exploration Corporation
SST — Supersonic Transport
SSTO — Single Stage To Orbit
STA — Space Transportation Association
STS — Space Tourism Society

T

t/Space — Transformational Space Inc

U

UN — United Nations
USAF — United States Air Force

V

V-2 — V-2 Vengeance rocket
VfR — Verein fur Raumschiffart
VSE — Vision for Space Exploration

W

WW — World War

X

X-1 — Bell X-1 rocket plane
X-15 — Northrop Grumman X-15 rocket plane
XCOR — Name of Jeff Greason's space tourism corporation
X-Prize — Organization dedicated to use of prizes to advance technology.

BIBLIOGRAPHY

Note that the numbering is arranged within each chapter, rather than as a continuum throughout the book.

AVIATORS CHAPTER

1. Wright, Orville, "How we Invented the Airplane", Dover Publications, originally published circa 1920.

2. Lindbergh, CA, "The Spirit of St Louis", Scribner – originally published 1953.

3. Schiff, Stacy, "Saint-Exupery a Biography", Henry Holt, 1994.

4. Goddard, Robert H, "Rockets", Dover Publications, originally published 1946.

5. Official Guide to the Smithsonian National Air and Space Museum, 2002.

6. Combs, Harry, "Kill Devil Hill – Discovering the Secret of the Wright Brothers", Houghton Mifflin, 1979.

7. Davies R.E.G., "Charles Lindbergh – an Airman, his Aircraft and his Great Flights", Palwadr Press, 1997.

ROCKET MEN CHAPTER

1. Tsiolkovsky, K.E., "Selected Works", Mir Publishers, Moscow 1968.

2. Harford, James, "Korolev", John Wiley and Sons, 1997.

3. Matson, Wayne R, (Editor), "Cosmonautics – a Colorful History", Cosmos Books, 1994.

4. Parkinson, Bob, "Interplanetary – A History of the BIS", British Interplanetary Society Publishing, 2008.

5. Von Braun, Wernher, "First Men to the Moon", Fischer Bucherei, 1958.

6. Clarke, Arthur C, "The Exploration of Space", Harper and Bros, 1951.

7. Buckbee, Ed and Schirra, Wally, "The Real Space Cowboys", Apogee Books, 2005.

8. Chaikin, Andrew and Kohl, Victoria, "Voices from the Moon", Viking Studio/Penguin Books, 2009.

9. Sachdev, D.K. (Editor), "Success Stories in Satellite Systems", AIAA Publishing, 2009.

10. Ordway, F, and Sharpe, M, "The Rocket Team", Apogee Books, 1979 and 2003.

X-MEN CHAPTER

1. Wolfe, Tom, "The Right Stuff", Farrar, Strauss and Giroux, 1979.

2. Yeager, C and Janos, L, "Yeager", Bantam Books 1985.

3. Crossfield, Scott, "Always Another Dawn", World Publishing, 1960.

4. Thompson, M.O., "At the Edge of Space", Smithsonian Books, 1992.

5. Jenkins, Dennis R, Landis, Tony R, "Hypersonic – the Story of the North American X-15", Speciality Press, 2003.

6. Tregaskis, Richard, "X-15 Diary: The Story of America's First Space Ship", University of Nebraska Press, 1961.

7. Wallace, Lane E, "Flights of Discovery", NASA History Office, 1996.

8. Pisano, D.A, Van der Linden, F.R., Winter, F.H., "Chuck Yeager and the Bell X-1: Breaking the Sound Barrier", Harry N Abrams, Inc., 2006.

PRESIDENTS CHAPTER

1. Wilson, Vincent, "The Book of the Presidents", American History Research Associates, 1993.

2. Dick, Steven J, Jacobs Robert, et al (Editors), "America in Space – NASA's First Fifty Years", Abrams, 2008.

ARTISTS CHAPTER

1. Szurovny, Geza, "The Art of the Airways", Zenith Press, 2002.

2. Parkinson, Bob, "Highroad to the Moon", British Interplanetary Society Publications, 1979.

3. Ordway, Frederik I, Editor, "Blueprint for Space", Smithsonian Institute Press, 1992.

4. Ordway, Frederick I, "The Rocket Team", Apogee Books, 2003.

5. Ley, Willy, "Rockets, Missiles and Space Travel, Viking Press, 1958.

6. Bradbury, Ray (Introduction), "The Art of Robert McCall", Bantam Books, 1992.

7. Hardy, David, "Visions of Space", Dragon's World, 1989.

8. Dean, James and Ulrich, Bertram, "NASA/Art 50 years of Exploration", Abrams, 2008.

9. Hartmann, WK, Sokolov A, et al, "In the Stream of Stars", Workman Publishing, 1990

10. Bean, Alan, "Apollo – an Eyewitness Account", Greenwich Workshop, 1998.

11. Scott, David, and Leonov, Alexei, "Two Sides of the Moon", St Martin's Press, 2004.

12. Launius, Roger and Ulrich, Bertram, "NASA and the Exploration of Space", Stewart, Tabori and Chang, 1998.

13. Kerrod, Robin, "NASA Visions of Space", Prion, 1990.

TOURISTS CHAPTER

1. Berinstein, Paula, "Making Space Happen", Plexus Publishing, 2002.

2. Spencer, John, "Space Tourism – Do you Want to Go?", Apogee Books, 2004.

3. Linehan, Dan, "SpaceShipOne- An Illustrated History", Zenith Press, 2008.

4. Anderson, Eric, "The Space Tourist's Handbook", Quirk Books, 2005.

5. Berns, Gregory, "Iconoclast", Harvard Business Press, 2008 (Particularly Chapter 7).

6. Belfiore, Michael, "Rocketeers", HarperCollins, 2007.

INDEX